INDUSTRIAL AUTOMATION AND PROCESS CONTROL

Jon Stenerson

Fox Valley Technical College

Prentice
Hall

Upper Saddle River, New Jersey
Columbus, Ohio

Library of Congress Cataloging-in-Publication Data

Stenerson, Jon
    Industrial automation and process control / Jon Stenerson.
      p. cm.
      Includes index.
      ISBN 0-13-033030-2

Editor in Chief: Stephen Helba
Editor: Ed Francis
Production Editor: Holly Shufeldt
Design Coordinator: Diane Ernsberger
Cover Designer: Thomas Borah
Cover photo: Corbis Stock Market
Production Manager: Brian Fox
Marketing Manager: Mark Marsden

This book was set in Times Roman by UG / GGS Information Services, Inc., and was printed and bound by R. R. Donnelley & Sons Company. The cover was printed by Phoenix Color Corp.

Pearson Education Ltd.
Pearson Education Australia Pty. Limited
Pearson Education Singapore Pte. Ltd.
Pearson Education North Asia Ltd.
Pearson Education Canada, Ltd.
Pearson Educación de Mexico, S.A. de C.V.
Pearson Education—Japan
Pearson Education Malaysia Pte. Ltd.
Pearson Education, *Upper Saddle River, New Jersey*

10 9 8 7 6 5 4 3 2
ISBN 0-13-033030-2

To my son, Jeff, of whom I am very proud

Preface

I began writing this text for my students when I was unable to find a practical text that covered the main topics in industrial automation. It was especially difficult to find practical and understandable information on process control. The other quandary I tried to address was that I wanted my students to study many topics in some detail but not in the depth that would require a whole textbook. I was also very fortunate to find some wonderful software (LogixPro) that emulates and simulates Rockwell Automation PLC programming. Luckily, I received permission from Mr. Bill Simpsom at **www.thelearningpit.com** to include a trial version in this text. This software will enable students to program and test their programs on graphics simulators in the software. Included are an I/O simulator and several other simulators for student work. The software will enable students to study and practice PLC programming in the comfort of their homes. The software is time limited. The full version may be purchased very inexpensively at any time. Note that a copy of Rockwell Automation instruction help has been included on the CD as have instructions for the use of LogixPro Programming and simulation software.

The book is organized as follows.

Chapter 1 is a quick overview of industrial automation. The basic devices and controllers employed in automated systems are presented to provide a framework for understanding material in the rest of the chapters.

Chapter 2 discusses safety and takes an in-depth look at lockout/tagout procedures for personnel who are working in industrial environments.

Chapters 3 through 6 provide the basic foundation for the use of PLCs. Chapter 3 focuses on the history and fundamentals of the PLC. Chapter 4 discusses coils, contacts, and the fundamentals of programming. This is the point at which students should begin to use the software included with the book as well as the Rockwell Automation instruction help. One of the more difficult things for students to understand is addressing. Chapter 5 is designed to enable the student to thoroughly understand this topic. Chapter 6 focuses on timers, counters, and logical program development.

Input/output modules and wiring are discussed in Chapter 7. The chapter discusses digital and analog modules. It stresses practical wiring of I/O modules. Other more specialized modules are also covered.

Chapter 8 considers arithmetic instructions and advanced instructions. The common arithmetic instructions, including add, subtract, multiply, divide, and compare, are covered. In addition, logical operators, average, standard deviation, trigonometric, and number system conversion instructions are covered. Sequencers and shift register programming are also covered.

Chapter 9 presents industrial sensors and their wiring. It focuses on types of sensors, including optical, inductive, capacitive, encoders, resolvers, ultrasonic, and thermocouples. The wiring and practical application of sensors is stressed.

Chapter 10 is an introduction to robots. Robot types, uses and applications are covered.

Chapter 11 covers the fundamentals of fluid power. The chapter begins with the basic theory of fluid power and then covers cylinders, valves, and simple applications.

Chapters 12 and 13 cover process control. Chapter 12 is a practical introduction to process and control and proportional, integral, and derivative (PID) control and tuning. Chapter 13 examines the types of process control devices and then applies them to practical level, flow, and temperature control systems.

Chapter 14 focuses on industrial communication. Communication devices will surely increase in importance as companies integrate their enterprises. This chapter provides a foundation for the integration of plant floor devices.

Chapter 15 presents the fundamentals of CNC programming. It begins by examining how CNC machines work. It examines closed-loop servo control and the basics of programming CNC machines. The chapter should give the student a good fundamental knowledge of how CNC machines work and how they are programmed.

Chapter 16 focuses on the installation, maintenance, and troubleshooting of PLC systems. It begins with a discussion of cabinets, wiring, grounding, and noise. The chapter provides the fundamental groundwork for the proper installation and troubleshooting of integrated systems.

Appendix A shows some common I/O device symbols.

I sincerely hope that the combination of topics covered in this book will provide you with the required depth and breadth you need in industrial automation.

In addition, particular thanks are due to Paul Brennan, Monroe Community College, Marvin Maziarz, Niagara County Community College, Robert J. Tuholski, University of Massachusetts—Lowell, and Charles Wiser, S.U.N.Y.—Alfred State College.

Contents

Overview of Industrial Automation

1

This chapter introduces the reader to the concepts of automated systems and the integration of various devices. It presents an overall perspective on what a system is, what components are involved, and how devices are integrated to provide a foundation for the rest of the topics in the book. All topics will be covered in greater detail in later chapters.

OBJECTIVES

Upon completion of this chapter you will be able to:
1. Explain terms such as *automated system, cell, control device, production device, support device,* feed back device, and *handshaking.*
2. Describe the differences between hard and flexible automation.
3. Describe the function and integration of devices such as robots, programmable logic controllers (PLCs), automated storage/automated retrieval systems (AS/AR), conveyors, hard automation, vision systems, and sensors.
4. Develop a flow diagram for a system.
5. Describe how devices are integrated into a cell.
6. Choose devices to produce a product and design a simple cell.
7. Evaluate an automation proposal based on payback period.

OVERVIEW OF INDUSTRIAL AUTOMATION

An *automated system* is a collection of devices working together to accomplish tasks or produce a product or family of products. An automobile, for example, is an automated system. The automobile has a brain box to receive inputs from various sensors and to control various outputs that regulate the engine's operation and other functions such as antilock braking.

A home burglar alarm system is another automated system. Its control box receives input from sensors and switches located on doors and windows of the house. If the control box receives a signal that a door or window has been opened, it sounds the alarm and calls the police department.

Industrial automated systems can be one machine or a group of machines called a *cell* (see Figure 1-1). Devices include those that actually produce the product and that provide support, control, and feedback to the system. The four basic types of devices in a cell are production, support, control, and feedback.

PRODUCTION DEVICES

Production devices may include robots, computer numerical control machines (CNCs), hard automation devices, and so on. Production devices add value to the product. They perform manufacturing processes such as machining, assembly, welding, painting, and other value-adding processes to form a completed part.

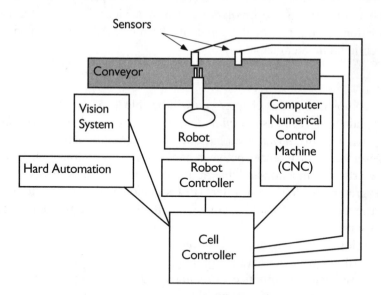

Figure 1–1 This figure shows a simple automated cell with several devices. The cell has a conveyor for bringing material into and out of the cell, a robot to move the material between devices in the cell, a CNC machine for machining the parts, a hard automation device for a special task, a vision system for inspecting the parts, some sensors for sensing parts, and a cell controller for integrating and controlling all of the other devices. Note that all the devices are connected to the cell controller.

Robots

Robots are used for many functions in a cell including repetitive ones such as moving and positioning parts between devices and production tasks such as welding parts together. Robots are very good for repetitive tasks. They are fast and accurate. Each of several types of robots—electric, pneumatic, and hydraulic—has its own advantages. Pneumatic robots are good for simple positioning tasks such as moving parts between devices. Pneumatic robots are inexpensive, fast, and accurate, but they are very limited in some ways. They are not good for complex tasks and are very limited in the number of positions to which they can move (see Figure 1-2). Electric robots, which are very versatile, are used for positioning, welding, and many other tasks. Electric robots are fast and accurate, but they are much more expensive than pneumatic robots. Hydraulic robots are good for painting and heavy applications. They are fast and can move very smoothly, making them effective for painting applications. They are also good for dangerous applications in which an electrical spark could cause an explosion.

Figure 1-2 This figure shows a programmer programming a Fanuc 6-axis robot with a teach pendant. (*Courtesy Hiromi Kugimiya.*)

Robot Integration   Robots are relatively easy to program and integrate into cells. Many robots are simply moved to each position and a key is used to teach each position. Other robots are programmed using specialized robot languages. Robots have digital (on/off) inputs and outputs (I/O) available. These inputs and outputs are easily connected to control devices such as PLCs, computers, or other devices. Most robot cells wait for an input to tell the robot what to do. It then performs a task and sends an output to a cell controller to tell the control device it has finished the task and is waiting for another (see Figure 1-3). This exchange of inputs and outputs between devices is called *handshaking*. Robots will be covered in greater detail in Chapter 10.

Computer Numerical Control Equipment (CNC)

CNC machines are machines whose motions and actions are controlled by a computer. They can be manufacturing processes, lathes and milling machines; metal-bending machines (press brakes); machines to punch metal; electrical discharge

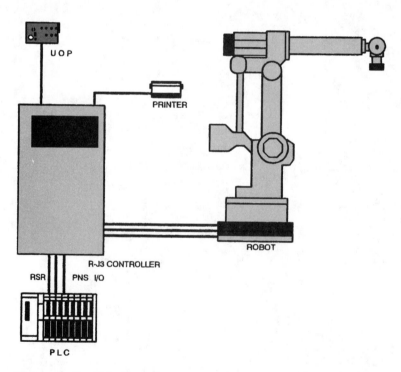

Figure 1–3   This figure shows how a robot communicates with a PLC. A digital output from the robot becomes an input to the PLC. A digital output from the PLC becomes an input to the robot. (*Courtesy Fanuc Robotics North America, Inc.*)

machines (EDMs), which use a wire or electrode and electrical sparks to cut metal; spring-making machines; and laser and plasma machines for cutting metal.

CNC machines are programmed with simple codes and numbers. The codes tell the machine what type of function to perform, and the numbers tell it how far to move. One example is N0025 M06 T5 (N0025 is a line number, M06 means change tools, and T5 means tool 5), which tells the CNC machine to change to tool number 5. CNC machines also have digital I/O available to talk to the control device. We will study CNC machines in greater detail in Chapter 15.

Hard Automation Devices for Production

A hard automation device is usually designed for one specific purpose and as a result, has little flexibility. When the product it produces is no longer manufactured, its hard automation device is normally worthless. Hard automation devices can be designed to be used as production devices or as support devices (see Figure 1-4).

SUPPORT EQUIPMENT

Support equipment may include automated storage/automated retrieval systems (AS/RS), conveyors, hard automation devices, and so on. AS/RS equipment is used for storage and retrieval of raw material, in-process products, and finished product. Computers control automated storage devices. The computer knows which areas are full and empty and decides where to store the product. The use of the computer also ensures that the oldest product in storage is used first. A typical AS/RS system has three axes of motion: an x-axis that moves along the floor, a y-axis for up and down motion, and a z-axis for moving in and out and putting the product in the storage area. Figure 1-5 shows an AR/AS system that has multiple rows of part storage and multiple conveyor lines to deliver parts to or from storage.

Conveyors

Conveyors are used to move the product between cells and processes and to sort product. This means that by using simple sensors or barcode readers, a conveyor can be used to send product to the right place by sensing which product is present.

Figure 1–4 Example of a hard automation device.

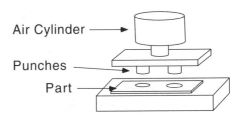

Figure 1–5   This automated storage/automated retrieval (AS/AR) system can put material anywhere in storage and keep track of it. (*Courtesy of White Systems, Inc.*)

Hard Automation Devices as Support Equipment

Hard automation equipment are generally designed to perform one task only. Hard automation devices are generally used only for very high production tasks. A hard automation device might be used to feed or align parts (see Figure 1-6). They are used when a flexible method would be more expensive. An example is a vibratory-bowl feeder, which is a container that is vibrated. The parts move up because of the vibrations and are aligned. Parts that are not aligned fall back into the bowl.

Palletizers

Palletizers are used to take boxes from a cell and pack them onto a pallet containing many boxes. The palletizer also wraps the boxes with plastic to hold them to-

Figure 1–6 This simple hard automation device was designed for one purpose only: to feed parts one at a time. When the air cylinder moves forward, one part is pushed out. This is a very simple device and needs to be controlled by some other device such as a PLC.

gether and protect them for shipping. A palletizer could be considered a hard automation device because it is usually just for one specific job. Robots that can be programmed to handle various products and various load types are more flexible palletizers.

CONTROL DEVICES

Programmable Logic Controllers (PLCs)

A PLC is the most common type of cell controller; it coordinates all other devices. The PLC is like the brain of the cell. It is a computer designed to be easy for electricians and technicians to use. It is programmed with a language called *ladder logic,* which is like the blueprints that electricians and technicians understand. A PLC is also useful in integrating and controlling other devices into systems. Chapters 3 through 8 will cover PLCs in greater detail.

Pneumatic and Hydraulic Logic

Simple systems with limited flexibility can be designed to be controlled by pneumatic or hydraulic logic. This will be studied in more detail in Chapter 10.

FEEDBACK DEVICES

Sensors

Sensors are available for almost every application. Sensors are like a cell's eyes, ears, and touch. They provide feedback to the control device to tell it what is happening in the cell. Sensors can be simple digital (on/off) types or more complex analog types that provide an output that is proportional to the input. An example is a sensor used to measure temperature. An analog sensor outputs a different current for every temperature and provides more information than a digital one but is more expensive.

Simple digital sensors are the most common. They include photo-sensors, inductive sensors used to sense metal objects, capacitive sensors used to sense any object, ultrasonic sensors used to sense objects or the distance to the object using sound, and simple limit switches. Sensors that contact the product are called *contact sensors*. Sensors that do not touch the product, such as photosensors, are called *noncontact*, or *proximity*, *sensors*. Electronic sensors (noncontact) are much more reliable and faster than mechanical sensors (contact sensors).

Vision Systems

Vision systems are used to inspect product quality (see Figure 1-7). The inspection might be used to check to ensure that all components are on a printed circuit board, to measure the size of product, or to read labels and ensure they are complete and accurate. Vision systems are increasing in use and importance in industry. They are much more accurate than humans for performing repetitive inspection tasks.

Vision System Operation   Vision systems look at contrast to make decisions. A picture taken with a camera is analyzed by the computer. Lighting of the object is very important because most vision systems can see only contrast. The vision processing board looks at each pixel to determine its brightness level, called a *gray level*. Typical systems have 256 levels of gray (brightness). The computer must look at all areas of the object and assign brightness levels, which represent contrast, to each.

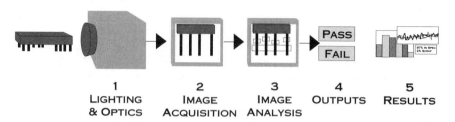

1	2	3	4	5
LIGHTING & OPTICS	IMAGE ACQUISITION	IMAGE ANALYSIS	OUTPUTS	RESULTS

Figure 1–7   A pictoral example of a vision system. On the left is the lighting and vision (1). The image acquisition hardware receives image information from the camera (2). Next the image must be processed (3). Based on the processing, outputs are turned on or off (4). The results from inspections can then be analyzed and used to adjust and improve the process (5). (*Courtesy PPT Vision Inc.*)

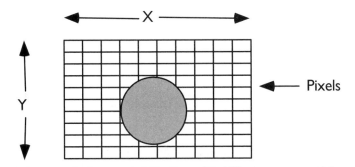

Figure 1–8 The matrix camera's image area is divided into small rectangles called *picture elements* or *pixels*. The horizontal axis is called the *x-axis*, and the vertical is called the *y-axis*.

Cameras The most common type of camera is a *CCD camera* (charge-coupled device). CCD cameras are sometimes called *RAM cameras* because they function similarly to dynamic RAM memory. A CCD camera is a matrix type camera (see Figure 1-8) that has a chip with an array of light sensors. The typical array contains a matrix of 520×520 (or more) pixels. The image is projected onto the chip by the lens, and each sensor looks at one small piece of the image (pixels or picture elements). Pixels are arranged in a rectangular pattern (see Figure 1-8), so the resolution will be better in one axis than the other. Remember that in our example there are 520×520 pixels, or 270,400 pixels, each of which must be analyzed. The vision processor must look at the brightness of each pixel and assign it a brightness level of between 0 and 255, a huge task. Then the vision system must analyze that information to make a decision.

Another type of camera is the CID (charge injection device). The CID camera, which can also be linear or matrix, has a charge injected into each cell in proportion to the intensity of the light striking it. The CID acts as computer memory and holds the charge long enough for the microprocessor that is scanning the matrix of cells to read it. CID is somewhat like static RAM.

The vision system does not need to analyze the whole image but only areas of interest, which can help speed up applications. These small areas are generally called *windows* or *regions of interest* (ROIs). The programmer draws these inspection areas on the screen as a rectangle. For example, if the application must measure a hole size, the operator can draw a rectangle around the hole. Then the system looks only at the pixels in that rectangle or ROI. Examine Figure 1-9. A camera image is shown on the left. The camera has taken a picture of a round part, which appears darker than the background, which appears to be white. The vision system assigns a brightness level to each pixel in the camera image. The graph on the right in Figure 1-9 shows a gray-level histogram of the frequencies

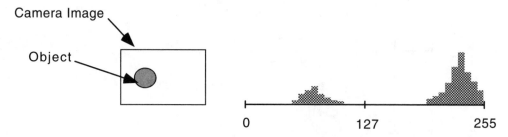

Figure 1–9   Camera image on the left and histogram on the right.

of pixel intensities. The pixels on the left of the histogram appear to be about 60 to 90 intensity. The pixels on the right of the graph are about 170 to 255 intensity. The pixels on the left must be the round image, and these on the right must be the background. Note that there are more bright (background) pixels than dark (object) pixels. Also note that while all of the background appears the same intensity to us, the system can see the differences in intensities.

Lighting   The most important consideration is usually the lighting. There are many light sources, fluorescent, incandescent, ultraviolet, and so on. Strobe lighting can be used to freeze the image when the parts are moving rapidly and need to be inspected. Backlighting is a technique using a translucent table (that has a light under it). The part is on the table and appears black because it blocks the light from below. Backlighting produces very good contrast, although front lighting is usually used.

There are several types of part analysis. One is called *blob analysis* (see Figure 1-10); it looks at the pixel gray levels and analyzes similar gray levels. It can look at the circumference of the part, area of the part, number and size of holes, and so on. By performing blob analysis, the vision system can determine which part is present, what its orientation is, and whether the part meets the specification. Vision systems use many other analysis techniques to analyze images.

Resolution   Resolution of the system is based on the number of pixels on the part image. In this case the round part covers 6.0 y pixels and 3.5 x pixels. If the part is 0.250 inches in diameter, the resolution in the y-axis would be 0.250 inches divided by 6 pixels, or 0.042 inches. The y-axis resolution is 0.250 inches divided by 3.5 pixels, or .071 inches.

A typical camera has an array of pixels of 520 × 520, or 270,400 pixels. This means that to maximize the part image over all of the pixels, the resolution is the part size divided by 520 for the x-axis and 520 divided by the y-axis. Remember that the y-axis resolution is finer because the pixels are rectangular in shape.

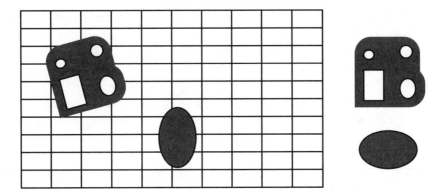

Figure 1–10 A simple example of blob analysis shows two blobs in the camera image. The actual parts are shown on the right side. The image processor looks at the outside circumference of the blobs, number of holes present, size of holes, distance between holes, area of the parts and holes, and so on. It then decides which part is which and rejects parts that do not match. Next it determines the position of the part in terms of xy pixels and the orientation. For example, the lower left corner of the left part is located at x1, y4, at an angle of 15 degrees. This information could be sent to a robot to move to the correct position and angle to pick up the part.

Applications   Because they are very versatile, vision systems have many applications. Vision systems are used to manufacture food and drugs, to check labels on products, and to check date and lot codes, which requires high-speed processing. Vision systems can be used in these ways:

- To measure sizes and part features.
- To check printed circuit boards or assembled products to ensure that all the components are present.
- To check all solder joints if special dye and lighting are used.
- To sort fruit, such as oranges, by size and quality.
- To check french fries to ensure no brown potato skin is left.
- To ensure properly packaged products.

Some most important factors to consider when purchasing a vision system are the speed required and the ease of use. Time is expensive, and a difficult-to-use system soon becomes very expensive.

Vision System Integration   Vision systems can communicate with other devices with digital I/O or through the computer. Digital I/O occurs by handshaking with a PLC. The PLC sends a bit pattern to the vision system to tell it which inspection program to run, and then the vision system runs the inspection and turns on one output if the part passed and a different output if the inspection failed. The PLC monitors the outputs from the vision processor.

Vision systems often use microcomputers, so they can exchange information with other software such as a database or spreadsheet. Dynamic data exchange (DDE) is one easy way for intelligent devices to exchange information in a Windows environment.

Programming a Vision System   The first and most crucial consideration in programming vision system applications is to start with a good image, which requires good lighting and contrast. The part or region to be inspected should fill as much of the image as possible to improve the resolution. The application will not work well with a poor image and will fail some good parts and accept some bad ones.

The newer vision systems have become much easier to use. Most now utilize Windows graphical interfaces to make programming easy.

Barcode

Barcoding is used extensively in business and industry; virtually every product in almost every store is barcoded now. Many barcode schemes are in use. One is universal product code (UPC), typically used for products in stores. Barcodes can carry product numbers, date information, and/or specifications. Print quality is important for properly reading barcodes. Barcoding technology is reasonably inexpensive to implement.

Radio Frequency (RF)

Radio frequency (RF) technology is increasing in use. It is more expensive than barcoding but is much more versatile. An RF system basically consists of a transmitter/receiver and a tag (see Figure 1-11); the tag stores the information. Two types of tags are active and passive. A *passive tag* contains no battery but receives power from the RF controller/transmitter. It is thus limited in distance capability. An *active tag* contains a battery. It has a longer read/write distance than a passive tag. Some tags are only readable; some are readable and writable. Tags can be as simple as one bit of storage or as complex as many thousand bytes of storage. Tags are somewhat expensive, so they are primarily used on more expensive, complex products or on pallets that carry parts through production. You have probably experienced situations in which a tag left on expensive merchan-

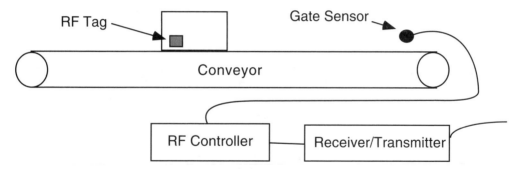

Figure 1–11   In a simple RF system, the part moves along the conveyor until it reaches the sensor called a *gate sensor*, which tells the RF controller when to read the tag. This ensures that it reads information from the correct product. The receiver/transmitter would be able to read or write to the tag.

dise in stores causes an alarm to go off when the tag passes the RF transmitter/receiver by the door.

DESIGN OF SYSTEMS

The first step when designing an automated system is to thoroughly understand the process. Will the cell be used to make just one type of product? How many products must be made per hour? Could it be used to make other parts? The design department can provide the information required to thoroughly understand the part requirements. The manufacturing department can explain its requirements.

Next, the devices needed are selected. Factors to consider include whether parts or raw material have to be supplied to the cell, how they leave the cell, what operations will be performed on each device, what type of controller to use, and where sensors are needed. Next, a drawing of the cell is made. It should show all devices that will be used in the cell. Finally, a sequence of operations is developed; it is a list of each step that occurs in the cell. It tells when and between which devices communication occurs.

The following is a short sample of a sequence of operations:

Part arrives on input conveyor: Sensor 1 (PLC input 1) senses the presence of the part.

PLC sees that sensor 1 is on and turns on PLC output 5 (robot input 1) to tell the robot to pick up the part.

Robot gets part and puts it in the CNC machine.

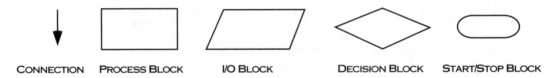

CONNECTION   PROCESS BLOCK   I/O BLOCK   DECISION BLOCK   START/STOP BLOCK

Figure 1–12   Common flow diagram symbols.

Robot turns on robot output 2 (PLC input 2) to tell the PLC that the part has been picked up.

PLC senses that there is a part in the CNC machine, sensor 3 (PLC input 3).

PLC turns on PLC outputs 5 and 6 to clamp the part in the fixture.

PLC turns on PLC output 8 to tell the CNC (CNC input 1) to machine the part.

CNC completes machining and turns on CNC output 1 (PLC input 4) to tell the PLC it is complete.

Another useful technique in system planning is a *flow diagram*, a pictorial representation of a system or its logic. Figure 1-12 shows several more common flow diagram symbols. The *decision* block answers a question, generally a yes or no question. It could perform a quality check or a piece count. The *process* block performs a process such as an arithmetic calculation, an assembly step, or other logical processing step. The *input/output* block inputs information into the system, and the *output* block outputs system information. The input might be from an operator or from the system; the output might be a printed report or a display on a terminal. Arrows are used to connect the blocks and to show the direction of operation.

Flow diagramming can facilitate the development of an application program development, particularly by breaking the application into small logical steps. Once the flow diagram is complete, it must be tested. This can be accomplished by applying some sample data and following it through the flow diagram. If the flow diagram works with sample data, the odds are that the logic is good. Once the flow diagram is complete, coding the application into ladder logic is a rather simple task. Figure 1-13 shows a heat treat application that was flow diagrammed. An explanation of the sequence of operations follows.

The operator inputs the number of parts to be hardened.

The parts are preheated to 1000 degrees Fahrenheit.

The parts are heated to 1650 degrees F for hardening.

The parts are quenched in oil to harden them.

The hardness is then checked; if it is incorrect, the parts are scrapped.

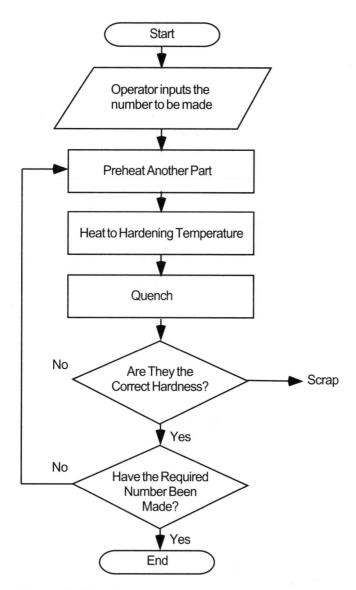

Figure 1–13 Flow diagram for a simple hardening process.

The piece count is checked to see whether enough parts have been hardened; if not, more are hardened.

When enough parts have been made, the process ends.

The second technique for developing systems or programs is called *pseudo-code*. It is an English representation of the system steps. The use of pseudocode requires the programmer to write logical steps for the system in descriptive statements or blocks. Pseudocode would not look much different from the steps of the flow diagram that was just developed. In fact, the two should be the same because the system should operate the same whether flow diagramming or pseudocode is used.

The designers next choose the devices and begin the actual design of the system. The costs and improvements can be estimated at this point. Automation should improve profitability and quality. After establishing the system's costs and benefits of the system, a cost justification can be prepared to determine whether to pursue the project.

Cost Justification

Automation is generally installed only when it saves money. Companies should avoid complex applications, especially for the first attempt at using a new technology. Determination of the payback period is often used to decide whether the project is a wise choice. Most companies want a payback period of much less than two years before investing in a robot. The following formula can be used to calculate payback for an application.

$$P = \frac{C}{W + I + D - (M + S)}$$

P = Payback period in years

C = Total cost of the system including installation

W = Annual wages of labor and fringe benefits for the workers who are replaced

I = Savings in terms of productivity, quality, materials, energy, etc.

D = Depreciation allowance

M = Expected system maintenance costs

S = Additional staffing for maintenance, programming, etc.

The formula is quite simple. Any cost that is a one-time expense for purchase, installation, or training goes above the line (cost).

Any yearly savings appear under the line minus any yearly additional costs. In essence, we are dividing the total cost of the system by the yearly savings. The answer will be the number of years it will take to pay for the system.

Cost Justification Problem   A company has designed a system to automate the loading and unloading of a die cast machine. The cost of the robot is $55,000; associated hardware costs $9,500. Other estimated costs are fixtures, $10,000; interfacing the robot with the machine, $6,500; maintenance and preventative maintenance each year, $15,000; and training for programer and maintenance, $8,000. The machine will run in two shifts, freeing two employees to do other work. The average cost of each worker, including fringe benefits, is $22,000. One worker will be retrained to do the programming, which is expected to cost $5,000 each year. The system is expected to result in savings per year of $12,000 due to improvements in productivity and quality. It is expected to be used for five years, so depreciation is calculated based on five years. What is the payback period? If the company requires a payback of 1.5 years or less, should the project be undertaken?

Solution to Cost Justification Problem   The depreciation is calculated first. The hardware costs equal $81,000 (55,000 + 9,500 + 10,000 + 6,500). The $8,000 one-time training costs are added to the hardware costs to get the total cost of the system. The $81,000 is divided by the 5 years of depreciation, resulting in annual depreciation of $16,200. The payback in this case is 1.7 years. Because the company wants a payback of 1.5 years or less, this project is not pursued.

$$\text{Payback} = \frac{89,000}{44,000 \ + \ 12,000 \ + \ 16,200 \ - \ (15,000 \ + \ 5,000)}$$

Other considerations may be important when deciding to pursue a project. The project might need to be undertaken because the work is hazardous or hiring qualified workers is difficult.

OPPORTUNITIES IN AUTOMATION

The rapid change in manufacturing has created a wealth of career opportunities. Skilled technicians and engineers who can work with these new technologies are in high demand. Automation has created a demand for designers who can utilize CAD/CAM systems to design and build automated systems and for people who understand how to program and integrate the devices. There is a need for skilled field engineers and maintenance people, and tremendous opportunities are available for sales application engineers who understand the new technologies and how to apply them.

The field of security biometrics, which is the use of technology for security, is exploding. Biometrics techniques include retina scanning, face recognition, and fingerprint scanning. Security systems are automated and integrated systems. Business and industry as well as airport security and immigration checks will experience a great need for technicians who understand and can effectively design and implement these systems. One current example is the use of industrial computer software, PLCs, and sensors for precision control systems. These systems utilize much of the same hardware and software that is used for automation and control in industry.

The wonderful thing is that these are highly paid and exciting careers. Skilled technicians and engineers are working with new technologies almost every day. The sky is the limit for a person with good work ethic and attitude in addition to a knowledge of automation.

QUESTIONS

1. Devices used in an automated system can have four purposes. What are they?
2. What is the difference between a hard automation device and flexible automation?
3. List at least four production devices and explain each.
4. Explain how robots can be integrated into a cell.
5. What is a hard automation device, and when is it used?
6. What is a PLC and what are its advantages?
7. What is a feedback device?
8. List three support devices and the purpose of each.
9. How is the resolution of a vision system determined?
10. Calculate the resolution for the following vision system. The actual part size is 0.150 inches.

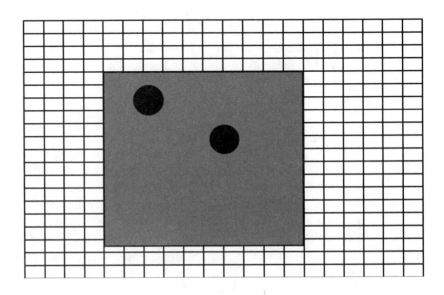

11. Assume a vision system with 520×520 resolution. How large an image (part) can we have if we need a resolution on the x-axis of at least 0.005 inch resolution?

12. Write a flow diagram for the following system. Assume that 100 parts have to be made and that each step is a separate diagram symbol.

    The operator puts a part in the heater and heats it to 1650 degrees for hardening.
    Has the part reached the required temperature? If it has, move to next step; if not, continue heating.
    Quenched the part in oil to harden it.
    Check the hardness; if it is incorrect, scrap the part. If it is correct, count it as a good part.
    Have enough parts been hardened? If not, harden more, if enough has been done, end the process.

13. Choose a simple part or product to manufacture and design a cell that would manufacture it. You must have devices to bring material into the cell and to take the product out of the cell. Draw a diagram complete with devices; use lines to represent the communications between devices and the cell controller. Write a sequence of operations that shows all communication and required feedback and a complete explanation of your cell. List all required devices and sensors.

14. A company has designed a system to automate the loading and unloading of a furnace. Related costs follow: robot, $75,000; associated hardware for the robot, $7,500; fixtures, $15,000; and interfacing the robot with the furnace, $5,500. It is estimated that no new maintenance personnel will be required but approximately $2,000 per month may be required for maintenance and preventive maintenance. No additional programming costs are expected. Training for programer and maintenance will cost $8,000. The furnace will run three shifts, freeing three employees to do other work. The average cost of each worker, including fringe benefits, is $32,000. The company has decided to depreciate the equipment over five years. It is expected to save $8,500 each year in productivity improvements. What is the payback period? If the company requires a payback of 24 or fewer months, should the project be undertaken?

15. A company has designed a system to automate an assembly operation. The costs are as follows: three robots, $105,000; a vision system, $7,500; three cameras, $1,500; associated hardware for the robot, $7,500; fixtures and hardware, $45,000. It is estimated that no new maintenance personnel will be required, but approximately $1,500 per month may be required for maintenance and preventive maintenance. Training will cost

approximately $12,000. The assembly cell will free up four of five employees to do other work. The average cost of each worker, including fringe benefits, is $27,000. The system's productivity and quality improvements should save the company $14,000 each year. The company believes that the system will be used for three years. What is the payback period? If the company requires a payback of 18 or fewer months, should the project be undertaken?

Safety

2

Safety should be considered the number one concern in the workplace, which has many dangers, including dangerous machinery and electrical hazards. Employees must always be aware of the dangers from which they must protect themselves and others. This chapter examines safe practices and lockout/tagout procedures.

OBJECTIVES

Upon completion of this chapter you will be able to:
1. Define what an accident is. Explain why accidents happen.
2. Describe how to prevent accidents.
3. Describe what lockout/tagout is and how it is used in industry.
4. Develop a lockout/tagout procedure.

ACCIDENT DEFINED

An *accident* is an unexpected action that results in injury to people or damage to property. Note that an accident is always unexpected. None of us ever think we will have an accident, but it happens to people just like us. An accident can easily result in the loss of an arm or leg, an eye, or even a life. Accidents cost enterprises huge amounts of money.

Most accidents are minor, but all are important. We should look for ways to prevent even minor accidents from recurring. A minor accident such as a cut could be more serious the next time. Every company is required to have procedures for reporting *all* accidents, even minor ones. Accidents are reported on forms so that the company can gather data in order to improve conditions and reduce the chance of future accidents and injuries. Companies often call these re-ports *incident report forms*.

CAUSES OF ACCIDENTS

Accidents can be prevented. Some of their causes follow.

1. Carelessness.
2. Use of wrong tools or defective tools or improperly using tools.
3. Unsafe work practices such as lifting heavy objects incorrectly.
4. Horseplay (playing around on the job).
5. Failure to follow safety rules or use safety equipment.
6. Inadequate equipment maintenance.

Items Commonly Sources of Accidents
1. Machines with moving or rotating parts.
2. Electrical machinery.
3. Equipment that uses high-pressure fluids.
4. Chemicals.
5. Sharp objects on machines or tools.

Remember that every piece of equipment can be dangerous!

ACCIDENT PREVENTION

Listed here are several ways to prevent accidents.

1. Design safety into the equipment. Place guards and interlocks on machines to prevent injury from moving parts and electrical shock. Provide lockouts on electrical panels so that an employee can lock the power out to repair equipment. Never remove guards or safety interlocks from machines while the machine is operating.
2. Use proper clothing, and eye and hearing protection including safety glasses. Employees in a noisy environment should wear ear protection. Never wear gloves, loose clothing, or jewelry around moving machinery.
3. Follow warning signs such as high voltage and danger signs. A yellow/black line means danger: Do not cross this line.
4. Follow safety procedures carefully.

SAFE USE OF LAB EQUIPMENT AND HAND TOOLS

Employees must follow procedures and learn good habits that will keep them safe on the job. Following is a list of safety guidelines.

1. Wear proper safety equipment, including eye glasses.
2. Know how to use the equipment. Do not operate equipment you do not understand.
3. Do not hurry.
4. Do not fool around.
5. Keep the work area clean.
6. Keep the floor dry.
7. Always beware of electricity.
8. Use adequate light.
9. Know first aid.
10. Handle tools carefully.
11. Keep tools sharp. A dull tool requires more pressure and is more likely to cause injury.
12. Use the proper tool.
13. Beware of fire or fumes. If you think you smell fire, stop immediately and investigate.
14. Use heavy extension cords. Thin extension cords can overheat, melting the insulation and exposing the wires or causing a fire.

OVERVIEW OF LOCKOUT/TAGOUT

On October 30, 1989, the Lockout/Tagout Standard, 29 CFR 1910.147 went into effect. It was released by the Department of Labor. Titled "The Control of Hazardous Energy Sources (Lockout/Tagout)," the standard was intended to reduce the number of deaths and injuries related to servicing and maintaining machines and equipment. Deaths and injuries resulting in tens of thousands of lost work days are attributable to maintenance and service activities each year.

Lockout is the placement of a lockout device on an energy-isolating device in accordance with an established procedure to ensure that the equipment being controlled cannot be operated until the lockout device is removed. *Tagout* is the placement of a tagout device on an energy-isolating device in accordance with an established procedure to indicate that the equipment being controlled may not be operated until the tagout device is removed. Only authorized employees who are performing the service or maintenance can perform tagout.

The lockout/tagout standard covers the servicing and maintenance of machines and equipment in which the unexpected start-up or energization of the machines or equipment or the release of stored energy could cause injury to employees. Machinery or equipment is considered to be energized if it is connected to an energy source or contains residual or stored energy. Stored energy can be found in pneumatic and hydraulic systems, springs, capacitors, and even gravity. Service and/or maintenance also includes activities such as constructing, installing, setting up, adjusting, inspecting, and modifying, machines or equipment. These activities include lubrication, cleaning or unjamming machines or equipment, and making adjustments or tool changes. The standard is intended to cover electrical, mechanical, hydraulic, chemical, nuclear, and thermal energy sources. The standard also establishes minimum standards for the control of such hazardous energy. The standard does not cover normal production operations, cords and plugs under exclusive control, and hot tap operations.

Hot tap operations involve transmission and distribution systems for substances such as gas, steam, water, or petroleum products in the repair, maintenance, and service activities that involve welding on a piece of equipment such as a pipeline, vessel, or tank, under pressure, to install connections or appurtenances. Hot tap procedures are commonly used to replace or add sections of pipeline without the interruption of service for air, gas, water, steam, and petrochemical distribution systems. The standard does not apply to hot taps when they are performed on pressurized pipelines provided that the employer demonstrates that continuity of service is essential, shut down of the system is impractical, documented procedures are followed, and special equipment is used to provide proven, effective protection for employees.

Employers are required to establish a program consisting of an energy control (lockout/tagout) procedure and employee training to ensure that before any employee performs any service or maintenance on a machine or equipment when the unexpected energizing, start-up or release of stored energy could occur and cause injury, the machine or equipment shall be isolated and rendered inoperative. The employer is also required to conduct periodic inspection of the energy control procedure at least annually to ensure that the procedures that were developed and the requirements of this standard are being followed. Only authorized employees may lock out machines or equipment. An authorized employee is one who has the authority and training to lock or tag out machines or equipment in order to perform service or maintenance on that machine or equipment.

Normal production operations, which include the utilization of a machine or equipment to perform its intended function, are excluded from lockout/tagout restrictions. Any work performed to prepare a machine or equipment to perform its normal production operation is called *setup*.

Another exclusion pertains to an employee working on cord and plug–connected electrical equipment for which exposure to unexpected energization or

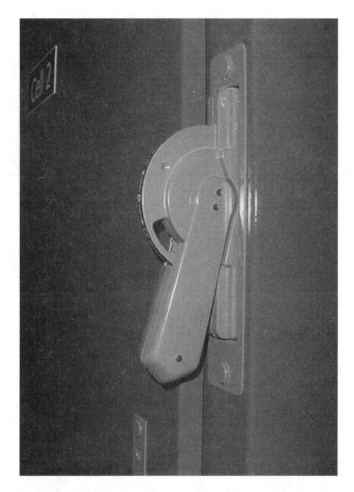

Figure 2–1 A typical electrical disconnect that is not locked out.

start-up of the equipment is controlled by unplugging the equipment from the energy source when the plug is under the exclusive control of the employee performing the service or maintenance.

An energy-isolating device is a mechanical device that physically prevents the transmission or release of energy. These devices include manually operated electrical circuit breakers; disconnect switches (see Figure 2-1), manually operated switches to disconnect the conductors of a circuit from all ungrounded supply conductors and to prevent independent operation of any pole; line valves (see Figure 2-2); and locks and any similar device used to block or isolate energy. Push buttons, selector switches, and other control circuit type devices are not energy-isolating devices. An energy-isolating device is capable of being locked

Figure 2–2 A typical pneumatic disconnect.

out if it has a hasp or other means of attachment to which, or through which, a lock can be affixed or a locking mechanism built into it. Other energy-isolating devices are capable of being locked out if this can be achieved without dismantling, rebuilding, or replacing the energy-isolating device or permanently altering its energy control capability.

New energy-isolating machines or equipment installed after January 2, 1990, must be designed to accept a lockout device.

Lockout Requirements

A lockout device utilizes a positive means such as a lock to hold an energy-isolating device in the safe position and prevent the energizing of a machine or equipment. A lock may be either a key or combination type. If a device is incapable of being locked out, the employer's energy control program shall utilize a tagout system.

Notification of Employees

The employer or an authorized employee must notify affected employees of the application and removal of lockout or tagout devices. Notification shall be given before the controls are applied to and after they are removed from the machine or equipment. *Affected employees* are defined as employees whose job requires operation or use of a machine or equipment on which service or maintenance is being performed under lockout or tagout or whose job requires work in an area in which such servicing or maintenance is being performed.

TAGOUT

A tagout device must warn against hazardous conditions if the machine or equipment is energized and must include a clear warning, such as, Do Not Start. Do Not Open. Do Not Close. Do Not Energize. Do Not Operate. A tagout device is a prominent warning device, such as a tag and a means to attach it, that can be securely fastened to an energy-isolating device in accordance with an established procedure to indicate that the equipment it controls may not be operated until the tagout device is removed by the employee who applied it. If the authorized employee who applied the lockout or tagout device is not available to remove it, specific procedures, which are given later, must be followed.

When used, tagout devices must be affixed in such a manner to clearly indicate that the operation or movement of energy-isolating devices from the "safe" or "off" position is prohibited. Tagout devices used with energy-isolating devices designed with the capability of being locked must be fastened at the same point at which the lock would have been attached. When a tag cannot be affixed directly to the energy-isolating device, it must be located as close as safely possible to the device in a position immediately obvious to anyone attempting to operate the device.

TRAINING

The employer must provide training to ensure that employees understand the purpose and function of the energy control program and have the knowledge and skills required for safely applying, using, and removing energy controls. Employees should be trained to:

Recognize hazardous energy sources.

Understand the type and magnitude of the energy available in the workplace.

Use methods necessary for energy isolation and control.

Understand the purpose and use of the lockout/tagout procedures.

All other employees whose work operations are or may be in an area where lockout/tagout procedures may be used shall be instructed about the prohibition of attempting to restart or reenergize machines or equipment that are locked out or tagged out.

When tagout procedures are used, employees must be taught the following:

Tags are warning devices and do not provide physical restraint on devices.

An attached tag is not to be removed without authorization of the person responsible for it, and it is never to be bypassed, ignored, or otherwise defeated.

Tags must be legible and understandable by all authorized employees, affected employees, and employees whose work operations are or may be in the area.

Tags may create a false sense of security. Their meaning needs to be understood by all.

Retraining

Retraining shall be provided for all authorized and affected employees whenever there is a change in their job assignments; in machines, equipment, or processes that present a new hazard; or in the energy control procedures. Additional retraining shall be offered when a periodic inspection reveals, or the employer has reason to believe, that an employee does not have adequate knowledge to use the energy control procedures. The retraining shall reestablish employee proficiency and introduce new or revised control methods and procedures as necessary. The employer shall verify that employee training has been accomplished and is being kept up to date. The verification shall contain each employee's name and dates of training.

Requirements for Lockout/Tagout Devices

Lockout and tagout devices must be singularly identified, the only device(s) used for controlling energy, and not used for other purposes. They must be durable, which means they must be capable of withstanding the environment to which they are exposed for the maximum period of time that exposure is expected. Tagout devices must be constructed and printed so that exposure to weather conditions, wet and damp locations, or acids and alkalis will not cause the tag to deteriorate or the message on the tag to become illegible (see Figure 2-3).

Lockout and tagout devices must be standardized within the facility in at least one of the following criteria: color, shape, or size. Print and format must also be standardized for tagout devices; they must be substantial enough to prevent removal without the use of excessive force or unusual techniques, such as the use of bolt cutters or other metal-cutting tools. Tagout devices, including their means of attachment, shall be substantial enough to prevent inadvertent or accidental removal. The devices must have a nonreusable type of attachment that is self-locking and nonreleasable with a minimum unlocking strength of at least 50 pounds. It should have the general design and basic characteristics equivalent to a one-piece, all-environment-tolerant nylon cable tie. They must be attached by hand. Lockout and tagout devices must identify the employee who applied them.

Application of Lockout/Tagout Devices   The established procedures for applying lockout or tagout procedures cover the following and shall be performed in the sequence indicated.

Figure 2–3 Typical tagout tag.

> # Warning
>
> This machine is locked out
>
> Reason - _____
> _____
> _____
> _____
> _____
> _____
>
> Name _____
> Date _____
> Time _____

1. Notify all affected employees that a lockout or tagout system will be used, and ensure that they understand the reason for the lockout. Before turning off a machine or equipment, the authorized employee must understand the type(s) and magnitude(s) of the energy, the hazard(s) of the energy to be controlled, and the method(s) to control the energy for the machine or equipment being serviced or maintained.

2. The machine or equipment shall be turned off or shut down using the procedures established for it. An orderly shutdown must be utilized to avoid any additional or increased hazards to employees as a result of the equipment stoppage.

3. All energy-isolating devices needed to control the energy to the machine or equipment shall be physically located and operated in such a manner as to isolate the machine or equipment from the energy sources.

4. Lockout or tagout devices shall be affixed to each energy-isolating device by authorized employees. Lockout devices, when used, shall be affixed in a manner to hold the devices in a "safe" or "off" position.

5. Following the application of lockout or tagout devices to energy-isolating devices, stored energy must be dissipated or restrained by methods such as repositioning, blocking, and bleeding down. If reaccumulation of stored energy to a hazardous level is possible, verification of isolation shall be continued until the service or maintenance is completed or until the possibility of such accumulation of energy no longer exists.

6. Prior to starting work on machine or equipment that have been locked out or tagged out, the authorized employee shall verify that it has actually been isolated and de-energized. This is done by operating the push button or other normal operating control(s) to ensure that the equipment will not operate.

The machine is now locked out or tagged out. Caution must be taken to ensure that operating controls are returned to the neutral or off position after the test. Before removing lockout or tagout devices and restoring energy to the machine or equipment, authorized employees shall follow procedures and take actions to ensure the following:

The work area has been inspected to ensure that nonessential items have been removed and that machine or equipment components are operationally intact.

The work area has been checked to ensure that all employees have been safely positioned or removed.

Affected employees shall be notified that the lockout or tagout devices have been removed.

Each lockout or tagout device must be removed from each energy-isolating device by the employee who applied it. The only exception to this is when the authorized employee who applied the lockout or tagout device is not available to remove it; the device then may be removed under the direction of the employer provided that specific procedures and training for such removal have been developed, documented, and incorporated into the employer's energy control program. The employer must demonstrate the following elements:

The authorized employee who applied the device was not available.

All reasonable efforts were made to contact the authorized employee to inform him or her that the lockout or tagout device has been removed before he or she resumes work at that facility.

Outside Personnel Working in the Plant   When outside service personnel (contractors, etc.) are engaged in activities covered by the lockout/tagout standard, the on-site employer and the outside employer must inform each other of their respective lockout or tagout procedures. The on-site employer must ensure that his employees understand and comply with the restrictions and prohibitions of the outside employer's energy control program.

Figure 2–4 Hasps allows multiple personnel to lock out machines or equipment.

Group Lockout or Tagout When a group of people perform service and/or maintenance, each person must use a procedure that protects him or her to the same degree that a personal lockout or tagout procedure would. The lockout/tagout standard specifies requirements for group procedures. Primary responsibility is vested in an authorized employee for a set number of employees. These employees work under the protection of a group lockout or tagout device (see Figure 2-4) that ensures that no one individual can start up or energize the machine or equipment. The authorized employee who is responsible for the group must ascertain the exposure status of individual group members with regard to the lockout or tagout of the machine or equipment. When more than one crew, craft, department, and so on is involved, overall job-associated lockout or tagout control responsibility is assigned to an authorized employee. This employee coordinates affected work forces and ensures continuity of protection. Each authorized employee must affix a personal lockout or tagout device to the group lockout device, group lockbox, hasp (see Figure 2-4), or comparable mechanism when she begins work and shall remove those devices when she stops working on the machine or equipment being serviced or maintained.

Shift or Personnel Changes Specific procedures must be utilized during shift or personnel changes to ensure the continuity of lockout or tagout protection between departing and arriving employees and to minimize exposure to hazards from the unexpected energization or start-up of the machine or equipment or the release of stored energy.

Tests of Machines, Equipment, or Components In some situations, lockout or tagout devices must be temporarily removed from the energy-isolating device and the machine or equipment energized to test or position the machine, equipment, or component. The following sequence of actions must be followed:

1. Clear the machine or equipment of tools and materials.
2. Remove employees from the machine or equipment area.

3. Remove the lockout or tagout devices as specified in the standard.
4. Energize and proceed with testing or positioning.
5. Deenergize all systems and reapply energy control measures in accordance with the standard to continue the servicing and/or maintenance.

SAMPLE LOCKOUT PROCEDURE

The following is a sample lockout procedure. Tagout procedures may be used when the energy-isolating devices are not lockable provided that the employer complies with the provisions of the standard that require additional training and more frequent and rigorous periodic inspections. When tagout is used and the energy-isolating devices are lockable, the employer must provide full employee protection and additional training and the more rigorous required periodic inspections. When more complex systems are involved, more comprehensive procedures may need to be developed, documented, and utilized.

Lockout Procedure for Machine 37

Note: This normally names the machine when multiple procedures exist. If only one exists, it is normally the company name.

Purpose

This procedure establishes the minimum requirements for the lockout of energy-isolating devices when maintenance or service is performed on machine 37. This procedure must be used to ensure that the machine is stopped, isolated from all potentially hazardous energy sources and locked out before employees perform any servicing or maintenance where the unexpected energization or start-up of the machine or equipment or release of stored energy could cause injury.

Employee Compliance

All employees are required to comply with the restrictions and limitations imposed upon them during the use of this lockout procedure. Authorized employees are required to perform the lockout in accordance with this procedure. All employees, upon observing a machine or piece of equipment that is locked out for service or maintenance, shall not attempt to start, energize, or use that machine or equipment.

A company may want to list actions to take in the event that an employee violates the procedure.

Lockout Sequence

1. Notify all affected employees that service or maintenance is required on the machine and that the machine must be shut down and locked out to perform the servicing or maintenance.

> The procedure should list the names and/or job titles of affected employees and how to notify them.

2. The authorized employee must refer to the company procedure to identify the type and magnitude of the energy that the machine utilizes, must understand the hazards of the energy, and must know the methods to control the energy.

> The types and magnitudes of energy, their hazards, and the methods to control the energy should be detailed here.

3. If the machine or equipment is operating, shut it down by the normal stopping procedure (depress the stop button, open switch, close valve, etc.).

> The types and locations of the operating controls of the affected machine or equipment should be detailed here.

4. Deactivate the energy-isolating devices so that the machine or equipment is isolated from the energy sources.

> The types and locations of energy-isolating devices should be detailed here.

5. Lock out the energy-isolating devices with assigned individual locks.

6. Stored or residual energy (such as that in capacitors, springs, elevated machine members, rotating flywheels, hydraulic systems, and air, gas, steam, or water pressure) must be dissipated or restrained by methods such as grounding, repositioning, blocking, bleeding down, and so on.

> The types of stored energy, as well as methods to dissipate or restrain the stored energy, should be detailed here.

7. Ensure that the equipment is disconnected from the energy sources by first checking that no personnel are exposed; then verify the equipment's isolation by operating the push button or other normal operating controls or by testing to ensure that the equipment will not operate. *Caution:* Return operating controls to the neutral or "off" position after verifying the isolation of the equipment.

The method of verifying the isolation of the equipment should be detailed here.

8. The machine or equipment is now locked out.

Returning the Machine or Equipment to Service

When the service or maintenance is completed and the machine or equipment is ready to return to normal operating condition, the following steps shall be taken:

1. Check the machine or equipment and the immediate area around it to ensure that nonessential items have been removed and that its components are operationally intact. Check the work area to ensure that all employees have been safely positioned or removed from the area.

2. After all tools have been removed from the machine or equipment, guards have been reinstalled, and employees are in the clear, remove all lockout or tagout devices. Verify that the controls are in neutral, and reenergize the machine or equipment. (*Note:* The removal of some forms of blocking may require reenergization of the machine before safe removal. Notify affected employees that the service or maintenance is completed and the machine or equipment is ready for use.)

Sample Lockout/Tagout Checklist

Notification   I have notified all affected employees that a lockout is required and the reason for the lockout.

Date _____   Time _____   Signature _____

Shutdown   I understand the reason the equipment is to be shut down following normal procedures.

Date _____   Time _____   Signature _____

Disconnection of Energy Sources   I operated the switches, valves, and other energy-isolating devices so that each energy source has been disconnected or isolated from the machinery or equipment. I have dissipated or restrained all stored energy such as springs, elevated machine members, capacitors, rotating flywheels, pneumatic and hydraulic systems, etc.

Date _____ Time _____ Signature _____

Lockout I have locked out the energy-isolating devices using my assigned individual locks.

Date _____ Time _____ Signature _____

Safety Check After ensuring that no personnel are exposed to hazards I have operated the start button and other normal operation controls to ensure that all energy sources have been disconnected and that the equipment will not operate.

Date _____ Time _____ Signature _____

The machine is now locked out.

QUESTIONS

1. What is an accident?
2. Explain three ways to prevent accidents.
3. Explain at least five things to do to be safe in a lab.
4. List the sources of energy typically found in an industrial environment.
5. What is an affected employee?
6. What is an authorized employee?
7. Define the term *lockout*.
8. Define the term *tagout*.
9. Describe the typical steps in a lockout/tagout procedure.
10. Write a lockout/tagout procedure for a cell that contains electrical and pneumatic energy.

chapter

Overview of Programmable Logic Controllers

3

In a very short period of time, programmable logic controllers (PLCs) have become an integral and invaluable tool in industry. In this chapter we examine how and why PLCs have gained such wide application and what a PLC is.

OBJECTIVES

Upon completion of this chapter you will be able to:
1. Explain some of the reasons that PLCs are replacing hardwired logic in industrial automation.
2. Explain such terms as *ladder logic, CPU, programmer, input devices,* and *output devices.*
3. Explain some of the features of a PLC that make it an easy tool for an electrician to use.
4. Explain how the PLC is protected from electrical noise.
5. Draw a block diagram of a PLC.
6. Explain the types of programming devices available for PLCs.

PLC COMPONENTS

The PLC is an industrial computer in which the hardware and software have been specifically adapted to the industrial environment and the electrical technician. Figure 3-1 shows the functional components of a typical PLC. Note the similarity to a computer.

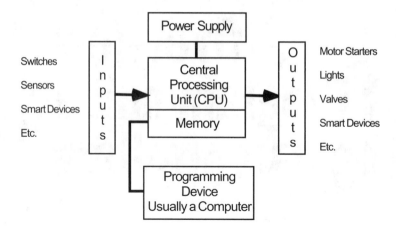

Figure 3–1   Block diagram of the typical components that make up a PLC. Note particularly the input section, the output section, and the central processing unit (CPU). The arrows from the input section to the CPU and from the CPU to the output section represent protection that is necessary to isolate the CPU from the real-world inputs and outputs. The programming unit is used to write the control program (ladder logic) for the CPU. It is also used to document the programmer's logic and troubleshooting the system.

Central Processing Unit

The central processing unit (CPU) containing one or more microprocessors is the brain of the PLC (see Figure 3-2). The CPU also handles the communication and interaction with the other components of the system. The CPU contains the same type of microprocessor in a microcomputer except that the PLC microprocessor program is written to accommodate ladder logic instead of other programming languages. The CPU executes the operating system, manages memory, monitors inputs, evaluates the user logic (ladder diagram), and turns on the appropriate outputs.

PLCs are hardened to be noise immune because factory floors are very noisy environments. Motors, motor starters, wiring, welding machines, and even fluorescent lights create electrical noise.

PLCs also have elaborate memory-checking routines to ensure that the PLC memory has not been corrupted by noise or other problems. Memory checking is a safety device that helps ensure that the PLC will not execute if its memory is corrupted for some reason. Most computers do not offer noise hardening or memory checking, but a few industrial ones do.

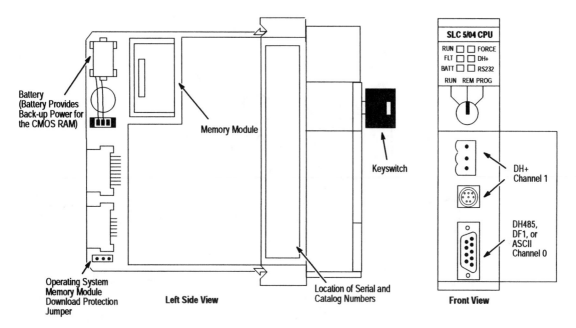

Figure 3–2 SLC processor module (CPU). Note the many indicators for error checking. Also note the keyswitch for switching modes run, rem (remote), and program. (*Courtesy Rockwell Automation Inc.*)

Memory

PLC memory can be of various types. Some of it is used to hold system memory, and some is used to hold user memory.

Operating System Memory

Read-Only Memory The PLC uses read-only memory (ROM) for the operating system. The operating system is burned into ROM by the PLC manufacturer. The operating system controls functions such as the system software that the user uses to program the PLC. The ladder logic that the programmer creates is a high-level language, which is a computer language that makes it easy for people to program. The system software must convert the electrician's ladder diagram (high-level language program) to instructions that the microprocessor can understand. ROM is nonvolatile memory, which means that even if the electricity is shut off, the memory is retained. The user cannot change ROM.

User Memory The memory of a PLC is broken into blocks that have specific functions. Some blocks are used to store the status of inputs and outputs. (*Note:*

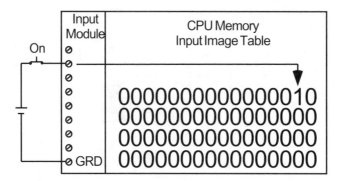

Figure 3–3 How the status of a real-world input becomes a 1 or a 0 in a word of memory. Each bit in the input image table represents the status of one real-world input.

Input/output is typically represented as I/O.) These are normally called *I/O image tables*. They keep the states of inputs and outputs. The real-world state of an input is stored as either a 1 or a 0 in a particular bit of memory. Each input or output has one corresponding bit in memory (see Figures 3-3 and 3-4). Other portions of the memory are used to store the contents of variables in a user program. For example, a timer or counter value would be stored in this portion of memory. Memory is also reserved for processor work areas.

Random Access Memory   Random access memory (RAM) is designed so that the user can read or write to the memory. It is commonly used for user memory. The user's program, timer/counter values, input/output status, and so on are stored in RAM.

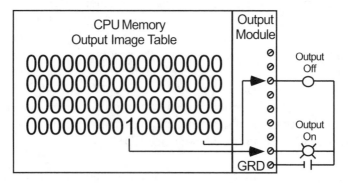

Figure 3–4 How a bit in memory controls one output. If the bit is a 1, the output is on. If the bit is a 0, the output is off. (This is active-high logic.)

RAM is volatile, which means that if the electricity is shut off, the data in memory is lost. This problem is solved by the use of a lithium battery, which takes over when the PLC is shut off. Most PLCs use CMOS-RAM technology for user memory. CMOS-RAM chips have very low current draw and can maintain memory with a lithium battery for an extended period of time, such as two to five years. A good preventive maintenance program should include a schedule to change batteries to avoid serious losses.

Figure 3-5 shows a battery replacement for a Rockwell SLC. SLC processors have a capacitor that provides at least 30 minutes of battery backup while the battery is being changed. The data in RAM is not lost if the battery is replaced within 30 minutes.

To replace the battery, the technician

Removes the battery from the retaining clips.

Inserts a new battery into the retaining clips.

Plugs the battery connector into the socket.

Reinstalls the module into the rack.

The battery in an SLC will last for approximately two years. The BATT LED on the front of the processor provides an alert when the battery voltage falls below a threshold level.

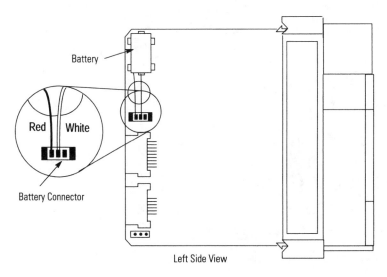

Figure 3–5 Replacement of a battery in a Rockwell SLC. (*Courtesy Rockwell Automation Inc.*)

Electrically Erasable Programmable Read-Only Memory   Electrically erasable programmable read-only memory (EEPROM) can function in almost the same manner as RAM. The EEPROM can be erased electrically. It is also nonvolatile memory, so it does not require battery backup. Available for many PLCs today, EEPROM modules are often small cartridges that can store several thousand bytes of memory.

PLC Programming Devices

A computer is the main device used to program a PLC. Handheld programmers are also available. Programming devices do not need to be attached to the PLC once the ladder has been written. These devices are used to write the user program for the PLC and to troubleshoot the PLC.

Handheld Programmers   Handheld programmers (see Figure 3-6), which are handy for troubleshooting, must be attached to a PLC to be used. Designed for use on the factory floor, they typically have membrane keypads that are immune to the contaminants in the factory environment. They can be carried easily to the

Figure 3–6   A Rockwell Automation handheld programmer. (*Courtesy Rockwell Automation Inc.*)

manufacturing system and plugged into the PLC. Once plugged in, they can be used to monitor the status of inputs, outputs, variables, counters, timers, and so on, eliminating the need to take a large programming device onto the factory floor. Handheld programmers can also be used to turn inputs and outputs on or off for troubleshooting. Turning I/O off or on by overriding the logic is called *forcing*. One disadvantage is that handheld programmers cannot show much of a ladder on the screen at one time.

Microcomputers

The microcomputer is the most commonly used programming device. It also can be used for off-line programming and storage of programs. One disk can hold many ladder diagrams. The microcomputer also can upload and download programs to a PLC and force inputs and outputs on and off.

This upload/download capability is vital for industry. Occasionally, PLC programs are modified on the factory floor to get a system up and running for a short period of time. It is vital that once the system has been repaired, the correct program be reloaded into the PLC. It is also useful to verify from time to time that the program in the PLC has not been modified. This can help to avoid dangerous situations on the factory floor. Some automobile manufacturers have set up communications networks that regularly verify the programs in PLCs to ensure that they are correct.

The microcomputer can also be used to document the PLC program. Notes for technicians can be added, and the ladder can be output to a printer for hard copy so that the technicians can study the ladder diagram.

RSLogix 500 is a Windows-based microcomputer software for programming Rockwell Automation PLCs (see Figure 3-7). RSLogix 500 is a powerful software package that allows off-line and on-line programming, the storage of ladder diagrams on a floppy disk, and uploading or downloading them from/to the PLC. The software can monitor the operation of the ladder while it is executing and can force the system inputs/outputs (I/O) on and off. This is extremely valuable for troubleshooting. It also allows the programmer to document the ladder. This documentation is invaluable for understanding and troubleshooting ladder diagrams. The programmer can add notes, names of input or output devices, and comments that may be useful for troubleshooting and maintenance. The addition of notes and comments enables any technician to readily understand the ladder diagram and to troubleshoot the system. The notes and/or comments could even specify replacement part numbers, which would facilitate rapid repair of any problems due to faulty parts. Previously, the person who developed the system had great job security because no one else could understand what had been done to it. A properly documented ladder allows any technician to understand it.

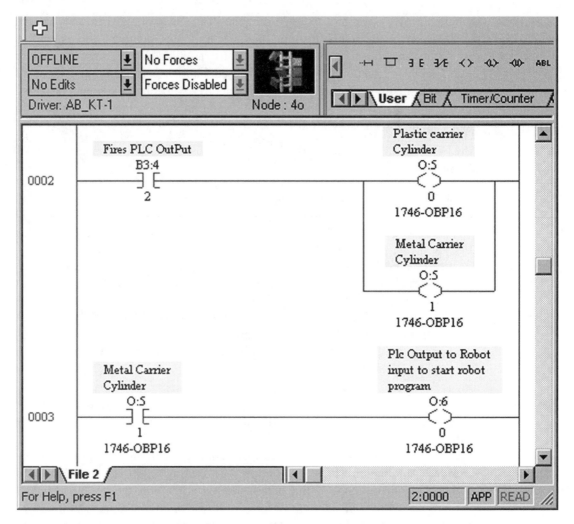

Figure 3–7 Typical PLC programming software. Note the comments and tag names to make the ladder logic more understandable. (*Courtesy Rockwell Automation Inc.*)

IEC 1131-3 Programming   The IEC has developed a standard for PLC programming. The latest IEC standard (IEC 1131-3) has attempted to merge PLC programming languages under one international standard. The standard allows PLCs now to be programmed in function block diagrams, instruction lists, C language, and structured text. The standard is accepted by an increasing number of suppliers and vendors of process control systems, safety-related systems, industrial personal computers, and so on. An increasing number of application software vendors offer products based on IEC 1131-3.

Power Supply

The power supply is used to supply power for the central processing unit. Most PLCs operate on 115 VAC. This means that the input voltage to the power supply is 115 VAC. The power supply provides various DC voltages for the PLC components and CPU. On some PLCs the power supply is a separate module, which is usually the case when extra racks are used. Each rack must have its own power supply.

The user must determine how much current will be drawn by the I/O modules to ensure that the power supply provides adequate current. Different types of modules draw different amounts of current. (*Note:* The PLC power supply is not typically used to power external inputs or outputs.) The user must provide separate power supplies to power the inputs and outputs of the PLC, although some smaller PLCs supply voltage to power the inputs.

Input Section

The input portion of the PLC performs two vital tasks, taking inputs from the outside world and protecting the CPU from the outside world. Inputs can be almost any device. The input module converts the real-world logic level to the logic level required by the CPU. For example, a 250-VAC input module would convert a 250-VAC input to a low-level DC signal for the CPU.

Common input devices include switches, sensors, and so on, often called *field devices*. Field devices are gaining extensive capability, especially the ability to communicate over industrial communications networks. Other smart devices, such as robots, computers, and even other PLCs, can act as inputs to the PLC.

The inputs are provided through the use of input modules. The user simply chooses input modules that meet the application's needs. These modules are installed in the PLC rack (see Figure 3-8).

The PLC rack serves several functions. It is used to physically hold the CPU, power supply, and I/O modules. The rack also provides the electrical connections and communications between the modules, power supply, and CPU through the backplane. Many PLCs allow the use of multiple racks. When more than one rack is used, the rack the I/O is in must be identified. Input/output numbering is a function of the slot into which the module is plugged. Rockwell Automation requires that the I/O number include the rack number.

The modules are plugged into slots on the rack (see Figure 3-9). This ability to plug modules in and out easily is one reason that PLCs are so popular. The ability to change modules quickly allows very rapid maintenance and repair.

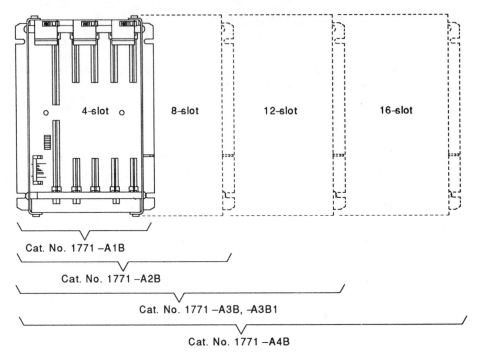

Figure 3–8   Various rack sizes. (*Courtesy Rockwell Automation Inc.*)

Figure 3–9   Modules, racks, and a rack filled with modules. (*Courtesy Rockwell Automation Inc.*)

Sometimes it is necessary to have more than one rack. Some medium-size and large PLCs allow more than one rack, and some applications have more I/O points than one rack can handle. At times it is desirable to locate some of the I/O away from the PLC. For example, imagine a very large machine. Rather than run wires from every input and output of the machine to the PLC, an extra rack might be used. The I/O on one end of the machine is wired to modules in the remote rack. The I/O on the other end of the machine is wired to the main rack. The two racks communicate with one set of wiring rather than running all of the wiring from one end to the other.

Optical Isolation The other task that the input section of a PLC performs is isolation. The PLC CPU must be protected from the outside world but be able to receive input data from it. Opto-isolation, short for "optical isolation," typically performs this function (see Figure 3-10). No electrical connection exists between the outside world and the CPU, and the two are separated optically. However, the outside world supplies a signal that turns on a light in the input card. The light shines on a receiver, and the receiver turns on.

The light separates the CPU from the outside world up to very high voltages. Even if there were a large surge of electricity, the CPU would be safe. (Of course, if the voltage is extremely large, the opto-isolator could fail and cause a circuit failure.) Optical isolation is used for inputs and outputs.

Input modules provide the user various troubleshooting aids. There are normally light-emitting diodes (LEDs) for each input. If the input is on, the CPU should see the input as a high (or a 1).

Input modules also provide circuits that *debounce* the input signal. Many input devices are mechanical and have contacts. When these devices close or open, unwanted "bounces" close and open the contacts. Debounce circuits make sure that the CPU sees only debounced signals. The debounce circuit also helps eliminate the possibility of electrical noise from firing the inputs.

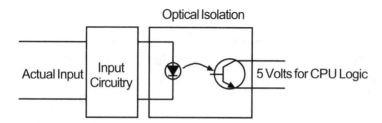

Figure 3–10 Typical optical isolation circuit. The arrow represents the fact that only light travels between the input circuitry and the CPU circuitry. There is no electrical connection.

Inputs to Input Modules   Sensors, which are commonly used as inputs to PLCs, can be purchased for a variety of purposes. They can sense part presence; count pieces; measure temperature, pressure, or size; sense for proper packaging; and so on. There are also sensors that are able to sense any type of material. Inductive sensors can sense ferrous metal objects, capacitive sensors can sense almost any material, and optical sensors can detect any type of material.

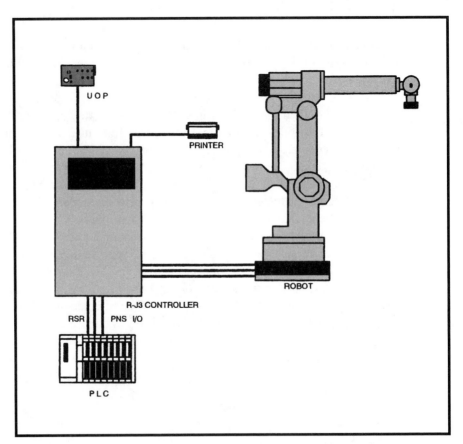

Figure 3–11 A "smart" device can also act as an input device to a PLC. The PLC can also output to the robot. Robots typically have a few digital outputs and inputs available for this purpose. The use of these inputs and outputs allows for some basic handshaking between devices. *Handshaking* means that the devices give each other permission to perform tasks at some times during execution to ensure proper performance and safety. When devices communicate with a digital signal, it is called *primitive communication*. (*Courtesy Fanuc Robotics North America Inc.*)

Other devices can also act as inputs to a PLC. Smart devices, such as robots, computers, and vision systems, often have the ability to send signals to a PLC's input modules (see Figure 3-11). These signals can be used for handshaking during operation. A robot, for example, can send the PLC an input when it has finished a program.

Sensor types and use are covered in detail in Chapter 9.

Output Section

The output section of the PLC provides the connection to real-world output devices. The output devices might be motor starters, lights, coils, valves, and so on. Often called *field devices*, they are gaining extensive capability, especially the ability to communicate over industrial communications networks. Field devices can be either input or output devices. Output modules can be purchased to handle DC or AC voltages and to output analog or digital signals. A digital output module acts as a switch whose output is either energized or deenergized. If the output is energized, it is turned on, as a switch would turn it on.

The analog output module is used to output an analog signal. For example, an analog module puts out a voltage that corresponds to the desired speed of a motor whose velocity we would like to control.

Output modules can be purchased with various output configurations of 8, 16, and 32 output modules. Modules with more than eight outputs, sometimes called *high-density modules*, are generally the same size as the eight-output modules but with many more components within the module. For that reason high-density modules do not handle as much current for each output because of the size of components and the heat they generate.

Current Ratings Module specifications list an overall current rating and an output current rating. For example, the specification may give each output a current limit of 1 ampere. If there are eight outputs, we would assume that the output module overall rating is 8A, but this is a poor leap in logic. The overall rating of the module current may be less than the total of the individuals. The overall rating might be 6A. The user must consider this when planning the system. Normally, each of the eight devices would not pull its 1A at the same time. Figure 3-12 shows an example of I/O wiring.

Output Image Table The output image table is a part of CPU memory (see Figure 3-13). The user's logic determines whether an output should be on or off. The CPU evaluates the user's ladder logic. If it determines that an output should be on, it stores a 1 in the bit that corresponds to that output. The 1 in the output

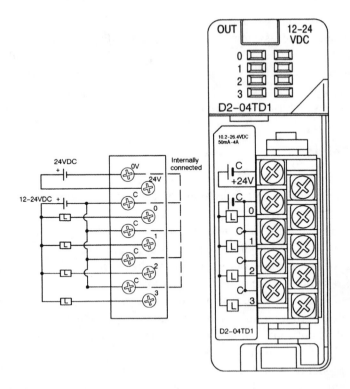

Figure 3–12 PLC Direct I/O Wiring. (*Courtesy PLC Direct.*)

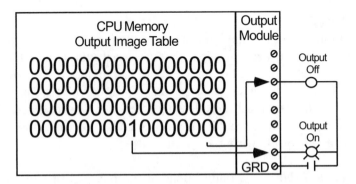

Figure 3–13 How a typical PLC handles outputs. The CPU memory contains a section called *the output image table*, which contains the desired states of all outputs. If there is a 1 in the bit that corresponds to the output, the output is turned on. If there is a 0, the output is turned off. This is called *active-high logic*.

Optical Isolation

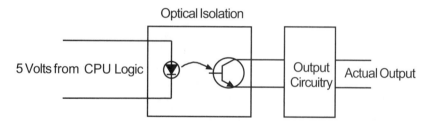

5 Volts from CPU Logic | Output Circuitry | Actual Output

Figure 3–14 How PLC output isolation works. The CPU provides a 5-volt signal that turns on the LED. The light from the LED is used to fire the base of the output transistor. There is no electrical connection between the CPU and the outside world.

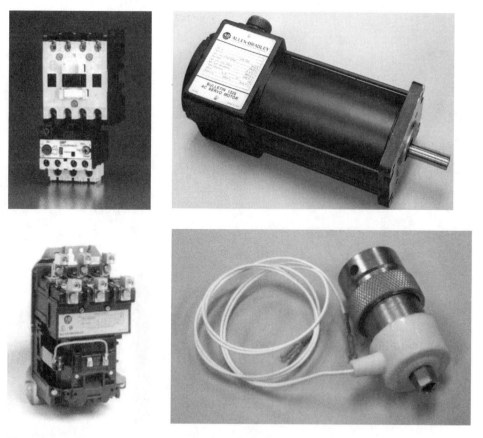

Figure 3–15 Several possible output devices including a contactor, AC motor, starter, and valve.

image table is used to turn on the actual output through an isolation circuit (see Figure 3-14).

The outputs of small PLCs are often relays. This allows the user to mix and match output voltages or types. For example, some outputs could then be AC and some DC. Relay output modules are also available for some of the larger PLCs. The other choices are transistors for DC outputs and triacs for AC outputs. Many types of field devices can be connected to outputs (see Figure 3-15 for a few examples).

PLC Applications

Programmable logic controllers are used for a wide variety of applications, such as to replace hardwired logic in older machines (see Figures 3-16 and 3-17). This can reduce the downtime and maintenance of older equipment. More importantly, PLCs

Figure 3–16 A PLC-controlled injection molding machine. (*Courtesy Rockwell Automation Inc.*)

Figure 3–17 Large flexographic printing press controlled by a PLC. Flexographic presses are used to produce packaging materials and other types of printed materials.

can increase the speed and capability of older equipment. Retrofitting an older piece of equipment with a PLC for control is much like getting a new machine.

PLCs are being used to control such processes as chemical production, paper production, steel production, and food processing. In processes such as these, PLCs are used to control temperature, pressure, mixture, concentration, and so forth. They are also used to control position and velocity in many kinds of production processes. For example, they can control complex automated storage and retrieval systems as well as equipment such as robots and production machining equipment.

Many small companies have recently been started to produce special-purpose equipment that is normally controlled by PLCs and that is very cost effective. Examples of this equipment are conveyors and palletizing, packaging, processing, and material handling. Without PLC technology, many small equipment design companies might not exist.

PLCs are being used extensively in position and velocity control. A PLC can control position and velocity much more quickly and accurately than can mechanical devices such as gears and cams. An electronic control system is not only faster but also does not wear out and lose accuracy as do mechanical devices.

PLCs are used for almost any process. Companies have PLC-equipped railroad cars that regrind and true rail track as they travel. PLCs also have been used to ring the perfect sequences of bells in church bell towers at exact times during the day and week. They are used in lumber mills to grade, size, and cut lumber for optimal output. The uses of PLCs are limited only by the imagination of the engineers and technicians who use them.

QUESTIONS

1. The PLC is programmed by technicians using:
   a. the C programming language
   b. ladder logic
   c. the choice of language used depends on the manufacturer
   d. none of the above

2. Changing relay control type circuits involves changing:
   a. the input circuit devices
   b. the voltage levels of most I/O
   c. the circuit wiring
   d. the input and output devices

3. The most common programming device for PLCs is the:
   a. dumb terminal
   b. dedicated programming terminal
   c. handheld programmer
   d. personal computer

4. CPU stands for central processing unit. True or false?

5. Opto-isolation:
   a. is used to protect the CPU from real-world inputs
   b. is not used in PLC but must be provided by the user
   c. is used to protect the CPU from real-world outputs
   d. both a and c

6. EEPROM is:
   a. electrically erasable memory
   b. electrically programmable RAM
   c. erased by exposing it to ultraviolet light
   d. programmable
   e. a, b, and d

7. RAM typically holds the operating system. True or false?

8. Typical program storage devices for ladder diagrams include:
   a. computer disks
   b. EEPROM
   c. static RAM cards
   d. all of the above
   e. none of the above

9. The IEC 1131 standard specifies characteristics for:
   a. PLC communications
   b. EEPROM
   c. memory
   d. PLC programming languages
   e. none of the above

10. Input devices would include the following:
    a. switches
    b. sensors
    c. other smart devices
    d. all of the above
    e. none of the above

11. Troubleshooting a PLC system:
    a. requires special PLC diagnostic equipment
    b. is much more difficult than troubleshooting relay type systems
    c. is easier because of indicators such as I/O indicators on I/O modules
    d. all of the above
    e. none of the above

12. Output modules can be purchased with which of the following output devices:
    a. transistor outputs
    b. triac outputs
    c. relay outputs
    d. all of the above
    e. both a and c

13. Field devices would include the following:
    a. switches
    b. sensors
    c. valves
    d. all of the above
    e. none of the above

14. If an output module's current rating is 1 A per output and there are eight outputs, the current rating for the module is 8 A. True or false? Explain your answer.

15. Describe how the status of real-world inputs are stored in PLC memory.

16. Describe how the status of real-world outputs are stored in PLC memory.

17. Define the term *debounce*. Why is it so important?

18. What is the difference between on-line and off-line programming?

19. What does it mean to force I/O?

chapter

Fundamentals of Programming 4

This chapter examines the basics of ladder logic programming, emphasizing terminology and common symbols, to enable the reader to learn to write basic ladder logic programs.

OBJECTIVES

Upon completion of this chapter you will be able to:
1. Describe the basic process of writing ladder logic.
2. Define such terms as *contact, coil, rung, scan, normally open,* and *normally closed.*
3. Write ladder logic for simple applications.

LADDER LOGIC

Programmable logic controllers (PLCs) are primarily programmed in ladder logic, which is really just a symbolic representation of an electrical circuit whose symbols were chosen to be similar to schematic symbols of electrical devices. This made it easy for people such as a plant electrician who has never seen a PLC to understand a ladder diagram and to learn to use the PLC. The main function of the PLC program is to control outputs based on the condition of inputs. The symbols used in ladder logic programming can be divided into two broad categories, contacts (inputs) and coils (outputs).

Contacts

Most inputs to a PLC are simple devices that are either on (true) or off (false). These inputs are sensors and switches that detect part presence, empty or full sta-

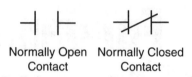

Figure 4–1   Normally open and normally closed contact.

Normally Open   Normally Closed
Contact   Contact

tus, and so on. See Figure 4-1 for the two common symbols for contacts. Contacts can be thought of as switches. The two basic kinds of switches are normally open and normally closed. A *normally open (XIC—examine if closed) switch* does not pass current until it is closed. A *normally closed (XIO—examine if open) switch* allows current flow until it is closed.

Think of a doorbell switch. Would you use a normally open switch or a normally closed switch for a doorbell? If you chose the normally closed switch, the bell would be on continuously until someone pushes the button on the switch. Pushing the button opens the contacts and stops current flow to the bell. The normally open switch is the correct choice. If the normally open switch is used, the bell does not sound until someone pushes the button.

Sensors detect for the presence of physical objects or quantities. For example, one type of sensor might be used to sense when a box moves down a conveyor, and a different type might be used to measure a physical property such as heat. Most sensors are switchlike: They are on or off, depending on what the sensor is sensing. Like switches, sensors that are either normally open or normally closed, can be purchased.

Imagine, for example, a sensor designed to sense a metal part as it passes the sensor. We could buy a normally open or a normally closed sensor for the application. If we wanted to notify the PLC every time a part passed the sensor, we could choose a normally open sensor. The sensor turns on only if a metal part passes in front of it and turns off again when the part has passed. The PLC could then count the number of times the normally open switch turned on (closed) and would know how many parts had passed the sensor. Normally closed sensors and switches are often used when safety is a concern. Sensors are covered in detail in a later chapter.

COILS

Contacts are input symbols; *coils* are output symbols. There are many types of real-world output devices: motors, lights, pumps, counters, timers, and relays. A coil is simply a representation of an output. The PLC examines the contacts (inputs) in the ladder and turns the coils (outputs) on or off, depending on the condition of the inputs. The basic coil is shown in Figure 4-2.

Figure 4–2 Ladder logic symbol for a coil.

LADDER DIAGRAMS

The basic ladder diagram looks similar to a ladder. It has two uprights and the rungs that make up the PLC ladder. The left and right uprights represent power. If we connect the left and right uprights through a load, power can flow through the rung from the left upright to the right upright.

Consider again the doorbell example with one input and one output. The PLC's ladder diagram has only one rung. See Figure 4-3. The real-world switch is connected to input 0 of the PLC. The bell also is connected to output 0 of the PLC (see Figure 4-4). The uprights represent a DC voltage used to power the doorbell. If the real-world doorbell switch is pressed, power can flow through the switch to the doorbell.

The PLC then runs the ladder and monitors the input continually and controls the output. This is called *scanning*. The amount of time it takes for the PLC to go through the ladder logic each time is called the *scan time*. Scan time varies from PLC to PLC. Even a slow PLC scan time is in milliseconds, so any PLC is fast enough. The longer the ladder logic is, the longer the scan time is.

Figure 4-5 illustrates the scan cycle. Note that this one rung of logic represents our entire ladder. Each time the PLC scans the doorbell ladder, it checks the state of the input switch *before* it enters the ladder (time 1). While in the ladder, the PLC then decides whether it needs to change the state of any outputs (evaluation during time 2). *After* the PLC finishes evaluating the logic (time 2), it turns on or off any outputs based on the evaluation (time 3). The PLC then returns to the top of the ladder, checks the inputs again, and repeats the entire process. The total of these three stages makes up scan time, which we discuss more completely later in this chapter.

Normally Closed Contacts

The normally closed contact will pass power until it is activated. A normally closed contact in a ladder diagram passes power while the real-world input associated with it is off.

Figure 4–3 Simple conceptual view of a ladder diagram.

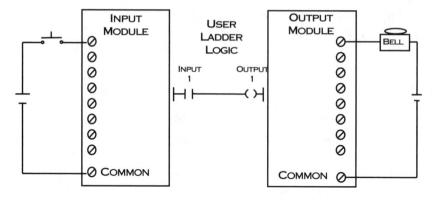

Figure 4–4 Conceptual view of a PLC system. The real-world inputs are attached to an input module (left side of the figure). Outputs are attached to an output module (right side of the figure). The center of the figure shows the logic that the CPU must evaluate by looking at the inputs and then turning on outputs based on the logic. In this case, if input 0 (a normally open switch) is closed, output 0 (the doorbell) turns on.

A home security system is an example of the use of normally closed logic. Assume that the security system was intended to monitor the two entrance doors to a house. One way to wire the house would be to wire one normally open switch from each door to the alarm, just like a doorbell switch (Figure 4-6). Then if a door opened, the system closes the switch and sounds the alarm. This works, but it has problems.

Assume that the switch fails. There are many ways in which it could fail: A wire could be cut accidentally, a connection could become loose, or a switch could break. The problem is that the homeowner would never know that the system was not working. An intruder could open the door; the switch would not work, and the alarm would not sound. Obviously, this is not a good way to design a system.

The security system should be set up so that the alarm sounds for an intruder and if a component fails. The homeowner surely wants to know when the system fails, which is far better than not sounding when an intruder enters.

Figure 4–5 A user's ladder logic is continually scanned.

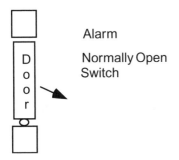

Alarm

Normally Open
Switch

Figure 4–6 Conceptual diagram of a home security system circuit. This is the wrong way to construct this type of application because the homeowner would never know when the system failed. The correct method is to use normally closed switches. The control system then monitors the circuit continuously to see whether the doors opened or a switch failed.

These considerations are even more important in an industrial setting where failure could cause an injury. The procedure of programming to ensure safety is called *fail safe*. The programmer must carefully design the system and ladder logic to ensure the safety of people and processes if a failure occurs. Consider Figure 4-7, which shows a robotic cell application. If the gate is opened, it opens the normally closed switch. The PLC sees that the switch has opened and sounds

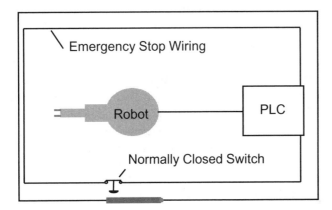

Figure 4–7 Cell application. This figure represents a robot cell with a fence with one gate. A PLC is used as a cell controller. The safety switch ensures that no one enters the cell while the robot is running. If someone enters the cell, the PLC senses that the switch opened and sounds an alarm. This case uses a normally closed switch. If the wiring or switch fails, the PLC thinks someone entered the cell and sounds an alarm. This system is fail safe.

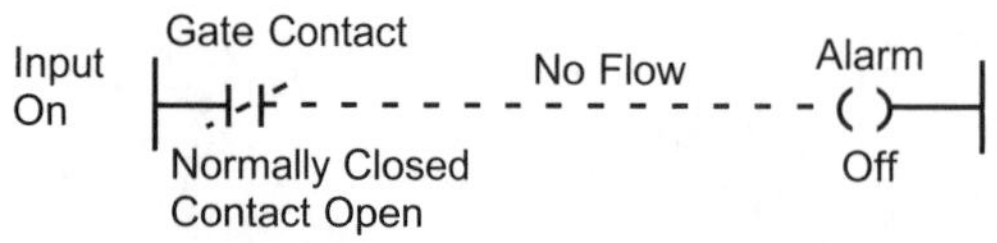

Figure 4–8 One rung of a ladder diagram. A normally closed contact is used in the ladder. If the switch associated with that contact is closed, it forces the normally closed contact open. No current flows to the output (the alarm). The alarm is off.

an alarm immediately to protect anyone who had entered the work cell. (In reality, we would sound an alarm and stop the robot to protect the intruder.)

The normally closed switch as used in ladder logic can be confusing. The normally closed contact in our ladder passes electricity if the input switch is off. (The alarm sounds if the switch in the cell gate opens.) The switch in the gate of the cell is a normally closed switch. (The switch in the cell normally allows electricity to flow.) If someone opens the gate, the normally closed gate switch opens, stopping electrical flow (see Figure 4-8). The PLC sees that there is no flow, the normally closed contact in the ladder allows electricity to flow, and the alarm is turned on.

Assume that a tow motor drives too close to the cell and cuts the wire that connects the gate safety switch to the PLC. What will happen? The alarm will sound because cutting the wire is similar to the gate opening the switch. It is a good thing that the alarm sounds if the wire is cut to warn the operator that something failed in the cell. The operator could then call maintenance and have the cell repaired. This is fail safe. Something in the cell failed, and the PLC shut down the system so that no one is hurt. The same would be true if the gate safety switch were to fail; the alarm would sound. If the switch were opened (someone opened the gate to the cell), the PLC would see that there is no power at the input (see Figure 4-9). The normally closed contact in the ladder logic is then closed, allowing electricity to flow. This causes the alarm to sound. Consider the rungs shown in Figure 4-10 and determine whether the output coils are on or off. (The answers follow.) Pay particular attention to the normally closed examples.

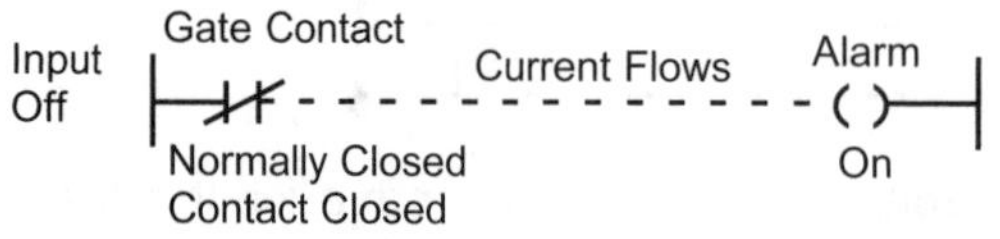

Figure 4–9 The input is off. (Someone opened the gate and opened the switch.) The normally closed contact is true when the input is false, so the alarm sounds. The same thing happens if a tow motor cut the wire that led to the safety sensor. The input is low and the alarm sounds.

Figure 4–10 Ladder
diagram exercise.

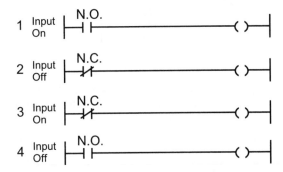

Answers to the ladder logic in Figure 4-10 follow.

1. The output in example 1 is on. The input associated with the normally
    open contact closes the normally closed contact in the ladder and passes
    power to the output.

2. The output in example 2 is on. The real-world input is off. The normally
    closed contact is closed because the input is off.

3. The output in example 3 is off. The real-world input is on, which forces
    the normally closed contact open (or the rung is false because the exam-
    ine-off input is on).

4. The output in example 4 is off. The real-world input is off so the
    normally open contact remains open and the output off.

Transitional Contacts

A *transitional contact* or *one-shot contact* is one type of special contact. The
symbol for this special contact is the normal contact symbol plus an arrow point-
ing either up or down (see Figure 4-11). A down arrow means that when this contact is energized, it will transition from high to low for one scan. An arrow point-
ing up means that when this type of contact is energized, it will transition from
off to on for one scan. They are called one shots because they are active for only
one scan when energized.

There are many reasons to use this type of contact such as to provide a pulse
for timing, counting, or sequencing. It is also used when an instruction needs to
be performed only once (not every scan). For example, if an instruction was used

—| ↑ |— —| ↓ |—

Figure 4–11 Two transitional contacts: low-to-high transitional contact and high-
to-low transitional contact.

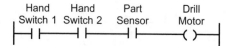

Hand Hand Part Drill
Switch 1 Switch 2 Sensor Motor

Figure 4–12 A series circuit. Hand switches 1 AND 2 AND the part sensor switch must be closed before the drill motor can be turned on to ensure that there is a part in the machine and that both of the operator's hands are in a safe location.

to add two numbers once, it would not be necessary to add them every scan. A transitional contact ensures that the instruction executes only on the desired transition.

Multiple Contacts

More than one contact can be put on the same rung. For example, think of a drill press. The engineer wants it to turn on only if there is a part present and the operator has one hand on each of the start switches (see Figures 4-12 and 4-13). This ensures that the operator's hands could not be in the press while it is running. This is equivalent to a series circuit or AND conditions. Hand switch 1 AND hand switch 2 AND the part sensor must be energized (on) for the drill motor to run.

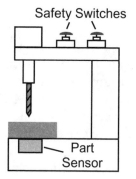

Figure 4–13 Simple drilling machine. The machine has two hand safety switches and one part sensor. Both hand switches and the part sensor must be true for the drill press to operate. This ensures that the operator's hands are not in the way of the drill. This is an AND condition. Switch 1 and switch 2 and the part sensor must be activated to make the machine operate. The ladder for the PLC is shown in Figure 4-14.

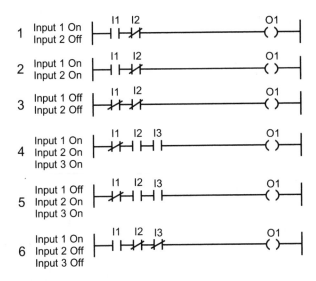

Figure 4-14   PLC ladder.

Note that the switches were programmed as normally open contacts. They are all on the same rung (series). All must be on for the output to turn on. If there is a part present and the operator's hands are on the start switches, the drill press will run.

If the operator removes one hand to wipe the sweat from her brow, the press stops. Contacts in series such as this can be thought of as logical AND conditions. In this case, the part presence switch AND the left-hand switch AND the right-hand switch would have to be closed to run the drill press.

Study the examples in Figure 4-14 and determine the status of the outputs. Answers to the ladder logic shown in Figure 4-14 are given below.

1. The output for rung 1 is on. Input 1 is on, which closes contact 1. Input 2 is off, so normally closed contact 2 is still closed. Both contacts are closed, so the output is on.

2. The output in rung 2 is off. Input 1 is on, which closes normally open contact 1. Input 2 is on, which forces normally closed contact 2 open. The output cannot be on because normally closed contact 2 is forced open.

3. The output in rung 3 is on. Inputs 1 and 2 are off so that normally closed contacts 1 and 2 remain closed.

4. The output in rung 4 is off. Input 1 is on, which forces normally closed contact 1 open.

5. The output in rung 5 is on. Input 1 is off so normally closed contact 1 remains closed. Inputs 2 and 3 are on, which forces normally open contacts 2 and 3 closed.

6. The output in rung 6 is on. Input 1 is on, forcing normally open contact 1 closed. Inputs 2 and 3 are off, which leaves normally closed contacts 2 and 3 closed.

BRANCHING

Often it is desirable to turn on an output for more than one condition. For example, in a house, the doorbell should sound under two conditions: the front button is pushed or the rear door button is pushed. The ladder, called a *branch*, is similar to Figure 4-15. Two paths (or conditions) can turn on the doorbell. (This can also be called a parallel condition or a *logical OR condition*.)

If the front door switch is closed, electricity can flow to the bell, or if the rear door switch is closed, electricity can flow through the bottom branch to the bell. Branching can be thought of as an OR situation. One branch OR another can control the output. ORs allow multiple conditions to control an output. This is very important in industrial control of systems. Think of a motor used to move a machine table. Two switches usually control table movement: a jog switch and a feed switch (see Figure 4-16). Both switches are used to turn on the same motor. This is an OR condition. The jog switch OR the feed switch can turn on the table feed motor. Evaluate the ladder logic shown in Figure 4-17.

These are the answers to the ladder logic in Figure 4-17.

1. The output in example 1 is off. Inputs 1 and 2 are off so output 3 is off.

2. The output in example 2 is off. Input 5 is on, which forces normally closed contact 5 open, so the output cannot be on whether or not input 1 OR input 2 is on. It should also be noted that these branching examples

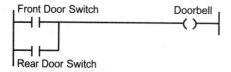

Figure 4–15   This figure shows a parallel condition. If the front door switch is closed, the doorbell sounds, OR if the rear door switch is closed, the doorbell sounds. These parallel conditions are also called *OR conditions*.

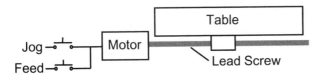

Figure 4–16 Conceptual drawing of a mill table. The two switches connected to the motor represent OR conditions. The jog switch or the feed switch can move the table.

have combinations of ANDs and ORs. In English, this example is input 1 AND input 5 OR input 2 AND input 5 will turn on output 12.

3. The output in example 3 is on. Input 2 is off, which leaves normally closed contact input 2 closed, AND input 5 is on, which closes normally open input 5, AND normally closed input 7 is off, which leaves normally closed contact 7 closed, which turns on output 1. This ladder has three OR conditions and combinations of ANDs.

4. In example 4, the output is on. Input 5 is off, which leaves normally closed contact 5 closed, AND input 1 is on, which forces normally open contact 1 closed, AND input 4 is off, which leaves normally closed

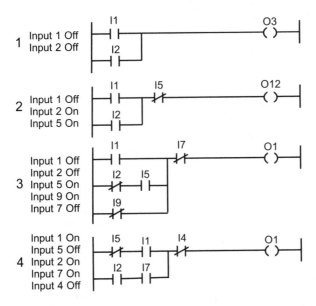

Figure 4–17 The text discusses the answers related to this figure.

Figure 4–18   Start/stop circuit.

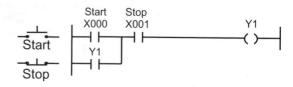

contact 4 closed, turning on the output. Inputs 2 and 7 are also both on, closing normally open contacts 2 AND 7.

Start/Stop Circuit

Start/stop circuits are very common in industry. Machines have a start button to begin a process and a stop button to shut off the system. Several important concepts can be learned from the simple logic of a start/stop circuit. Examine Figure 4-18. Notice that the actual start switch is a normally open push button. Pressing it closes the switch. Releasing the button opens the switch. The stop switch is a normally closed switch. When pressed, it opens. Note also that the stop switch contact in the ladder logic is programmed as a normally open switch. It is normally energized in operation because the real-world switch is normally closed.

Now examine the ladder. When the start switch is momentarily pressed, power passes through X000. Power also passes through X001 because the real-world stop switch is a normally closed switch. The output (Y1) is turned on. Note that Y1 is then also used as an input on the second line of logic. Output Y1 is on, so contact Y1 also closes. This is called *latching*. The output latches itself on even if the start switch opens. Output Y1 shuts off only if the normally closed stop switch (X001) is pressed. If X001 opens, then Y1 is turned off. The system requires the start button to be pushed to restart the system. Note that the real-world stop switch is a normally closed switch, but that in the ladder, it is programmed normally open for safety.

There are as many ways to program start/stop circuits (or ladder diagrams in general) as there are programmers. Figures 4-19 and 4-20 show examples of the wiring of start/stop circuits, for which safety is always the main consideration.

Figure 4–19   Start/stop circuit.

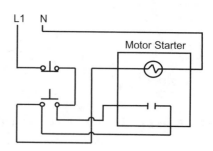

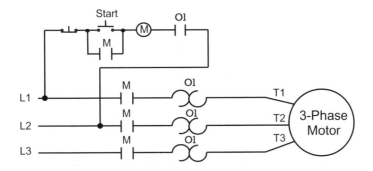

Figure 4–20 Start/stop circuit.

PLC SCANNING AND SCAN TIME

Now that you are familiar with some basic PLC instructions and programming, it is important to understand the way a PLC executes a ladder diagram. Most people would like to believe that a ladder is a very sequential thing, we like to think of a ladder as first things first. We would like to believe that the first rung is evaluated and acted on before the next, and so on. We would like to believe that the CPU looks at the first rung, goes out and checks the actual inputs for their present state, comes back, immediately turns on or off the actual output for that rung, and then evaluates the next rung, but this is not exactly the way it happens. Misunderstanding the way the PLC scans a ladder can result in programming bugs.

Scan time can be divided into two components: *I/O scan* and *program scan*. When the PLC enters run mode, it first takes care of the I/O scan (see Figure 4-21). The I/O scan can be divided into the *output step* and the *input step*. During these two steps, the CPU transfers data from the output image table to the output modules (output step) and then from the input modules to the input image table (input step).

The third step in the I/O scan is *logic evaluation*. The CPU uses the conditions from the image table to evaluate the ladder logic. If a rung is true, the CPU writes a 1 into the corresponding bit in the output image table. If the rung is false, the CPU writes a 0 into the corresponding bit in the output image table. Note that nothing concerning real-world I/O is occurring during the evaluation phase. This is often a point of confusion. The CPU is basing its decisions on the states of the inputs as they existed before it entered the evaluation phase. We would like to believe that if an input condition changes while the CPU is in the evaluation phase, it would use the new state, but cannot. (*Note:* It actually can if special instructions called *immediate update contacts* and *coils* are used. For the most part, most ladders do not utilize immediate in-

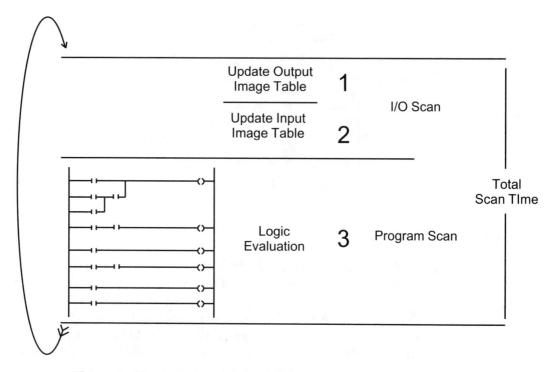

Figure 4–21 Generic example of PLC scanning.

structions. This is covered in more detail later.) The states of all inputs were frozen before it entered the evaluation phase. The CPU does not turn on/off outputs during this phase either, which is only for evaluation and updating the output image table status.

Once the CPU has evaluated the entire ladder, it performs the I/O scan again. During the I/O scan, the output states of real-world outputs are changed, depending on the output image table. The real-world input states are then transferred again to the input image table.

All of this takes only a few milliseconds (or less). PLCs are very fast, which is why troubleshooting can be so difficult. Scan time is the sum of the times it takes to execute all of the individual instructions in the ladder. Simple contacts and coils take very little time. Complex math statements and other types of instructions take much more time. Even a long ladder diagram normally executes in less than 50 milliseconds or so. Different PLC brands and models have considerably different speeds. Manufacturers normally give scan time in terms of fractions of milliseconds per K of memory, which gives a rough idea of the scan times of various brands.

QUESTIONS

1. What is a contact? A coil?

2. What is a transitional contact?

3. For what are transitional contacts used?

4. Explain the term *normally open* (XIC—examine if closed).

5. Explain the term *normally closed* (XIO—examine if open).

6. What are some uses of normally open contacts?

7. Explain the terms *true* and *false* as they apply to contacts in ladder logic.

8. Design a ladder that shows series input (AND logic). Use X5, X6, AND NOT (normally closed contact) X9 for the inputs and Y10 for the output.

9. Design a ladder that has parallel input (OR logic). Use X2 and X7 for the contacts.

10. Design a ladder that has three inputs and one output. The input logic should be X1 AND NOT X2, OR X3. Use X1, X2, and X3 for the input numbers and Y1 for the output.

11. Design a three-input ladder that uses AND logic and OR logic. The input logic should be X1 OR X3, AND NOT X2. Use contacts X1, X2, and X3. Use Y12 for the output coil.

12. Design a ladder in which coil Y5 will latch itself in. The input contact should be X1. The unlatch contact should be X2.

13. Draw a diagram and thoroughly explain what occurs during a PLC scan.

14. Examine the following rungs and determine whether the output for each is on or off. The input conditions shown represent the states of real-world inputs.

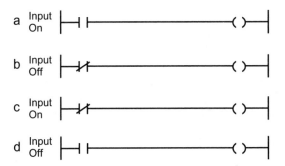

15. Examine the following rungs and determine whether the output for each is on or off. The input conditions shown represent the states of real-world inputs.

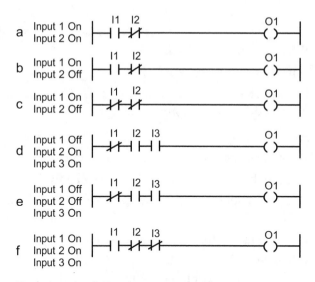

16. Examine the following rungs and determine whether the output for each is on or off. The input conditions shown represent the states of real-world inputs.

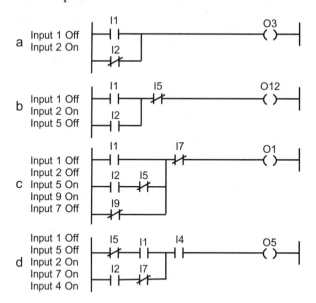

PROGRAMMING PRACTICE

RSLogix Relay Logic Instructions

These exercises utilize the LogixPro software included with this book. To install the software, simply insert the CD into your CD-ROM drive; it should self-install. Note that the software will operate only for a limited trial period. After that (or anytime before) you may register by paying a small fee to have the full version of the software.

Description of the Software

This is a complete working version of LogixPro. In the trial or evaluation mode, only the Silo, Door, and Traffic simulations are fully available for user programming. In addition, until LogixPro has been registered, File Save and Printing functionality is disabled. For those unfamilar with RSLogix, a program file for the Silo Simulation is included to get you started. When you have LogixPro running, select the Silo Simulation, then click Load in the File Menu, and select the file "silo.rsl". Once it is loaded and you can see it, you can then "DownLoad" it to the PLC. At this point you can place the PLC into the "RUN" mode. If all goes well, just clicking on the simulations START push button should start the whole process. We do not support online editing, so remember to place the PLC in the "PGM" mode to edit and then "Download" to the PLC before attempting to "RUN" again.

If you need help with the RSLogix addressing or instructions, you can go to two places. With your mouse, select Start, then Programs, then TheLearningPit, then LogixPro, and then you may choose the Instruction Set Reference, LogixPro, the readme file, or the student exercises. You can also log onto the Internet and go to the "LogixPro . . . Student Exercises and Documentation" page entry, which is listed on TheLearningPit.com home page. Also remember to try clicking on rungs, instructions, and so on with the right mouse button to locate popup editing menus and so forth.

First Exercise

This exercise is designed to familiarize you with the operation of LogixPro and to walk you through the process of creating, editing, and testing simple PLC programs utilizing the relay logic instructions supported by RSLogix. This is a very important exercise because it will prepare you for programming the more complex applications in later chapters. It will also enable you to achieve basic competancy in the use of Rockwell Automation RSLogix software. Go to the Student Exercises as described and complete the Relay Logic—Introductory Exercise.

chapter

Rockwell Automation Addressing and Instructions

This chapter examines Rockwell Automation memory addressing and instructions. It provides information regarding how to write basic ladder logic programs.

OBJECTIVES

Upon completion of this chapter, you will be able to:

1. Explain Rockwell Automation memory organization.
2. Explain Rockwell Automation addressing.
3. Define addresses for various types of files.
4. Explain the use of various Rockwell Automation instructions.
5. Write programs that utilize Rockwell Automation instructions.

UNDERSTANDING ROCKWELL FILE ORGANIZATION AND ADDRESSING

Organization and addressing Rockwell files can be one of the more confusing topics when learning how to program a PLC, so it is very important that you study this material carefully. This will dramatically reduce the frustration you experience as you begin to write programs.

First, Rockwell Automation divides its memory system into two types: program and data (see Figure 5-1). Remembering that there are two types of memory is very important. A mental image of two separate memory areas—program and data—might help. The program area of memory has 256 files. The data area has 256 files.

There are two types of memory in a Rockwell Automation PLC: program and data. The program area of memory has 256 files (file 0 through file 255). The data area of memory has 256 files (file 0 through file 255). Lets look at program memory first.

System and Program Files   Data Files

0   1   2   255   0   1   2   255

Figure 5–1   Two types of memory in a Rockwell Automation PLC.

Rockwell File Organization

Program File Memory   Program memory has 256 files (file 0 through file 255). Program files contain controller information, the main ladder program, and any subroutine programs (see Figure 5-2). The memory program file types follow:

File 0 contains various system-related information and user-programmed information such as processor type, I/O configuration, processor file name, and password.

File 1 is reserved.

File 2 contains the main ladder diagram.

Files 3–255 are user created and accessed according to subroutine instructions residing in the main ladder program file. In other words, the user can break the ladder diagram into logical portions of the total application program. Each portion of the ladder diagram can then be accessed as needed from the main program in file 2. For example, the main ladder diagram in file 2 could contain the logic for the user to choose manual or automatic operation. If the user chooses automatic operation, the logic in file 3 (or whatever number the programmer desired) is executed. If the user chooses manual operation, the logic in file 4 (or whatever number the programmer desired) is executed.

Data Memory   Data memory for Rockwell PLCs also has 256 files available. Data files are the files used to write ladder logic. It may be helpful to imagine an

System and Program Files

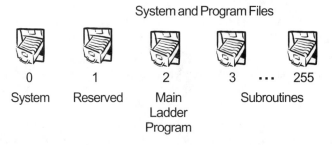

0   1   2   3   ...   255
System   Reserved   Main   Subroutines
                     Ladder
                     Program

Figure 5–2   Program files.

office where information is stored in files. Imagine 256 files (see Figure 5-1), each with a specific use.

Data files contain the status information associated with external I/O and all other instructions used in the main and subroutine ladder program files. In addition, these files store information concerning processor operation. The files can also store "recipes" and lookup tables if needed.

Data files are organized by the type of data they contain. The data file types follow:

The first file (file 0) is used to store the status of the outputs of the PLC. If the status of an output needs to be changed, put a 1 or 0 in the correct bit in file 0.

File Number	Type	Use
0	Output	This file stores the states of output terminals for the controller.
1	Input	This file stores the states of input terminals for the controller.
2	Status	This file stores the controllers's operation information. This file can be very useful for troubleshooting the controller and the program operation.
3	Bit	This file can be used for internal relay bit storage
4	Timer	This file stores the accumulated value, preset value and status bits for timers.
5	Counter	This file stores the accumulated value, preset value and status bits for counters.
6	Control	This file stores the length, pointer position, and status bits for specific instructions such as sequencers and shift registers.
7	Integer	This file can be used to store integer numbers or bit information.
8	Floating Point	This file can be used to store single precision non-extended 32-bit float numbers.
9-255	User Defined	These files can be used for any of the previously defined types by the user. Note that the whole file number would have to be used for the same type.

Figure 5–3 File types and their uses.

File 1 is used to store the status of inputs (see Figure 5-3). To check the status of inputs, look in file 1.

File 2 is reserved for PLC status information. This file contains status information that could be very helpful for troubleshooting and program operation.

File 3 stores *bit* information. Bits can be very helpful in logic programming. They can be used to store information about conditions or as contacts or coils for nonreal-world I/O.

File 4 is used for timer information.

File 5 is reserved for counter information.

File 6 is reserved for control. It is used when working with shift registers and or sequencers.

File 7 stores integers (whole numbers).

File 8 stores floating-point numbers (decimal numbers). This is the last files reserved for special uses.

File 9 through file 255 are user configurable. The user can use them for any purpose, such as more room to store integers. File types may not be mixed however; the entire file must be reserved for integer use.

Figure 5-3 shows the use of the files. Take some time to study these types and their uses. It will help you understand addressing.

Rockwell File Addressing

I/O addressing is relatively straightforward. See Figure 5-4 for the identifiers and numbers for file types. Study the example in Figure 5-5. The first letter of the input address is called an *identifier*. In this case, it is an input, so the identifier is an *I*, meaning that this is an input. The second item in the address is a 1. This is the file number. The default number for input files is 1. The next item is a colon, which is a delimiter. A *delimiter* defines what will come next. If the delimiter is a colon, the next item to follow is always the slot number of the module if we are naming inputs or outputs. Figure 5-6 shows an example of a PLC rack of modules. In Figure 5-5, this input address is in slot 2. The next item is a slash, which is also a delimiter. The slash delimiter always means the next item is a bit number. In this case, this is input 3. So, to review, this is input 3 on the input module that is in slot 2.

Study Figure 5-7, which shows an example of output addressing. The first item in the address is an *O* for output. The next item is the default file number for outputs (file 0). The next item, a colon, is a delimiter, meaning that the next item to follow is the slot number. In this case it is the output module in slot 4. The next

File Type	Identifier	File Number
Output	O	0
Input	I	1
Status	S	2
Bit	B	3
Timer	T	4
Counter	C	5
Control	R	6
Integer	N	7
Float	F	8

Figure 5–4 Default file identifiers and numbers.

Figure 5–5 Address name of input 3 on the input module in slot 2.

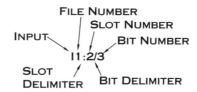

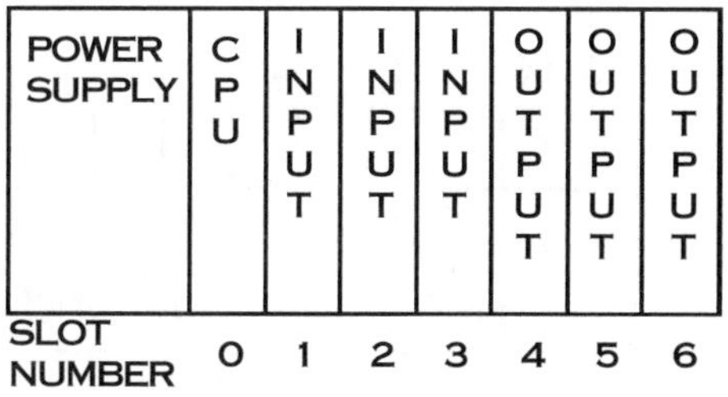

Figure 5–6 Note the input modules in slots 1, 2, and 3, and the output modules in slots 4, 5, and 6.

POWER SUPPLY	CPU	INPUT	INPUT	INPUT	OUTPUT	OUTPUT	OUTPUT
SLOT NUMBER	0	1	2	3	4	5	6

Figure 5–7 Address name of output 6 on the output module in slot 4.

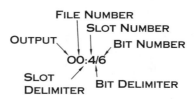

O0:5/12—Output 12 in slot 5, output file 0
O:6/7—Output 7 in slot 6, output file 0 (default file number 0)
I1:2/8—Input 8 in slot 2, input file 1
I:3/5—Input 5 in slot 3, input file 1 (default file number 1)
O0:4/12—Output 12 in slot 4, output file 0
O:5/1—Output 1 in slot 5, output file 0 (default file number 0)
I:1/9—Input 9 in slot 1, input file 1 (default file number 1)

Figure 5–8   Examples of I/O addresses.

Figure 5–9   Address name of bit element 64, file 3, bit 12.

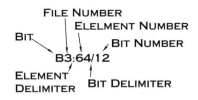

Figure 5–10   Address name of counter C, file 5, counter 0.

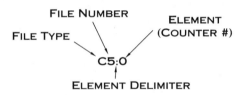

15	14	13	12	11	10	9	8	7	6	5	4	3	2	1	0	Element
																0 C5:0
PRE/LEN																1 C5:0.PRE
ACC/POS																2 C5:0.ACC

Figure 5–11   Example of counter memory.

Figure 5–12   Address name of timer 7, file 4, accumulated time.

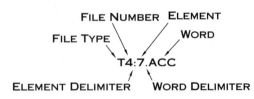

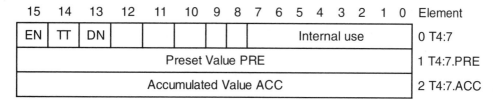

15	14	13	12	11	10	9	8	7	6	5	4	3	2	1	0	Element
EN	TT	DN							Internal use							0 T4:7
Preset Value PRE																1 T4:7.PRE
Accumulated Value ACC																2 T4:7.ACC

Figure 5–13 Example of timer memory.

item is the delimiter. The slash means that the number that follows is the bit number (output number), in this case, output 6 on the output module in slot 4. Figure 5-8 has several examples of I/O addresses. Make sure you understand them.

Figure 5-9 is an example of a bit element address. In this example, the type identifier (*B*) specifies a bit type. The file number is *3*, the element within the file is *64*, and the bit number is *12*.

Elements for timers, counters, control, and ASCII files consist of three words. Figure 5-10 is an example of a counter address. In this case, *C* stands for counter, *5* is the file number, and this counter is 0. Figure 5-11 shows that counters occupy three words of memory. The word *PRE* in C5:0.PRE holds the preset value, and *C5:0.ACC* holds the accumulated value. Timers and counters are covered in more detail in Chapter 6.

Figure 5-12 is an element address of T4:7.ACC. The *T* identifier stands for a timer, the file number is 4, the colon is the delimiter for timer *7*, and *ACC* specifies the accumulated value word. Note that timers utilize three words of memory just as counters do (see Figure 5-13). Timers will be covered in depth in Chapter 6.

Figures 5-14 and 5-15 show examples of an integer element address. In Figure 5-15 the *N* stands for integer, *7* is the file number, and *12* identifies the element.

Floating-point files have two-word elements. Figure 5-16 shows an example of floating-point element address for F8:0. Note that two words hold the floating-point number. In this example, the *F* stands for floating point, the *8* is the file number, and the *0* means the first two words in file 8. Remember that each floating-point number requires two words of storage.

String files have 42-word elements. Figure 5-17 shows an example. Note the use of the decimal point delimiter to specify the particular word in the string file.

Figure 5–14 Address name of integer element for file type N, number 7, element 12.

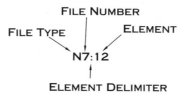

FILE NUMBER

FILE TYPE ELEMENT

N7:12

ELEMENT DELIMITER

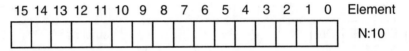

Figure 5-15 contents:

```
15 14 13 12 11 10 9  8  7  6  5  4  3  2  1  0    Element
                                                  N:10
```

Figure 5–15   Example of integer memory.

```
15  14  13  12  11  10  9  8  7  6  5  4  3  2  1  0    Element
                                                        F8:0
```

Figure 5–16   Example of floating-point element address.

```
15 14 13 12 11 10 9  8  7  6  5  4  3  2  1  0    Element
                                                  STf:0.LEN
                                                  STf:0.1

              *                      *
              *                      *
              *                      *

                                                  STf:0.41
```

Figure 5–17   Example of string file element address.

USER-DEFINED FILES

Users can define files for any purpose. File numbers 9–255 are available for any purpose for the user. An entire file must have the same use, however; for example, if the user desires to use file 10 for additional integers, file 10 can be used for integers only. See Figure 5-18 for an example of user-defined files 9–255.

User-Defined Files		
File Type	**Identifier**	**File Number**
Bit	B	
Timer	T	
Counter	C	
Control	R	
Integer	N	9-255
Float	F	
String	St	
ASCII	A	

Figure 5–18 User-defined file number 9–255.

ROCKWELL AUTOMATION CONTACTS

Examine If Closed

Rockwell Automation calls its normally open instruction *examine-if-closed* instruction contact. If a real-world input device is on, this type of instruction (XIC) is true and passes power (see Figure 5-19). If the input bit from the input image table associated with this instruction is a 1, the instruction is true. If the bit in the input image table is a 0, the instruction associated with this particular input bit is false.

Examine If Open

Rockwell Automation calls its normally closed contacts *examine-if-open contacts.* The examine-if-open instruction (XIO) can also be called a *normally closed instruction.* This instruction responds in a manner opposite to the normally open in-

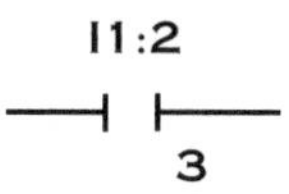

```
     I1:2
 ──┤ ├──
     3
```

Figure 5–19 Examine-if-closed instruction (XIC). If the CPU sees an on condition at bit I1:2/3, this instruction is true. The numbering of the input instruction is as follows: This is an input in file 1 located on an input module in slot 2 and is real-world input number 3 of the input module. The numbering of inputs and outputs are covered later.

Figure 5–20   Examine-if-open instruction (normally closed).

I1:2

—]/[—

3

struction. If the bit associated with this instruction is a 0 (off), the instruction is true and passes power. If the bit associated with the instruction is a 1 (true), the instruction is false and does not allow power flow.

Figure 5-20 shows an examine-if-open instruction (XIO). The input is number 3 on a input module in slot 2. If the bit associated with I1:2/3 is true (1), the instruction is false (open) and does not allow power flow. If bit I0:2/3 is false (0), the instruction is true (closed) and allows power flow.

SPECIAL CONTACTS

Many special-purpose contacts are available to the programmer, although the original PLCs did not have many. Sharp programmers used normally open and normally closed contacts in ingenious ways to turn outputs on for one scan, to latch outputs on, and so on. PLC manufacturers added special contacts to their ladder programming languages to meet these needs. The programmer can now accomplish these special tasks with one contact instead of a few lines of logic.

Immediate Instructions

Immediate instructions are used when the input or output being controlled is very time dependent. For example, for safety reasons we may have to update the status of a particular input every few milliseconds. If our ladder diagram is 10 milliseconds long, the scan time is too slow. This could be dangerous. The use of immediate instructions allows inputs to be updated immediately as they are encountered in the ladder. The same is true of output coils.

Rockwell Automation SLC 500 Immediate Input with Mask

The immediate input instruction (IIM) is used to acquire the present state of one word of inputs (see Figure 5-21). Normally, the CPU must finish all evaluation of

Figure 5–21   Immediate input instruction (IIM).

Figure 5–22 Immediate input instruction (IIM) addressing format.

I:2	INPUTS OF SLOT 2, WORD 0
I:2.1	INPUTS OF SLOT 2, WORD 1
I:1	INPUTS OF SLOT 1, WORD 0

the ladder logic and then update the output and input image tables. In this case, when the CPU encounters this instruction during ladder evaluation, it interrupts the scan. Data from a specified I/O slot is transferred through *a mask* to the input data file. A mask is a binary number that the programmer specifies in the instruction. If the bit in the mask is a 1, the input condition is transferred to the input data file. If the bit is a zero, it is not transferred. This makes the data available to instructions following the IIM instruction. This means that it gets the real-time states of the actual inputs at that time and puts them in that word of the input image table. The CPU then returns to evaluating the logic using the new states it acquired. This is used only when time is a crucial factor. Normally, the few milliseconds a scan takes is fast enough for anything we do, but in some cases the scan time is too long for some I/O updates. Motion control is one case. Updates on speed and position may be required every few milliseconds or less. We cannot safely wait for the scan to finish to update the I/O. In these cases immediate instructions are used.

Figure 5-22 shows an immediate input instruction (IIM) addressing format. When the CPU encounters the instruction during evaluation, it immediately suspends what it is doing (evaluating) and updates the input image word associated with I/O slot 2, word 0.

Rockwell Automation One-Shot Rising Instruction

The one-shot rising instruction (OSR) is a retentive input instruction that can trigger an event to happen only one time (see Figure 5-23). When the rung conditions that precede the OSR go from false to true, the OSR instruction is true for one scan. After the one scan, the OSR instruction becomes false even if the preceding rung conditions that precede it remain true. The OSR becomes true again only if the rung conditions preceding it make a transition from false to true. Only one OSR can be used per rung.

Figure 5–23 Rockwell Automation OSR instruction.

ROCKWELL AUTOMATION COILS

Output Energize

The output energize instruction (OTE) is the normal output instruction that sets a bit in memory. If the logic in its rung is true, the output bit is set to a 1. If the logic of its rung is false, the output bit is reset to a zero. Figure 5-24 shows an output energize (OTE) instruction. This particular example is real-world output 1 of the output module in slot 5. If the logic of the rung leading to this output instruction is true, output bit O:5/1 is set to a 1 (true). If the rung is false, the output bit is set to a 0 (false).

Rockwell Automation Immediate Outputs

The immediate output instruction (IOM) is used to update output states immediately. In some applications, the ladder scan time is longer than the needed update time for certain outputs. For example, an output not turned on or off before an entire scan is complete might cause a safety problem. In these cases, and when performance requires immediate response, immediate outputs (IOMs) are used. When the CPU encounters an immediate instruction, it exits the scan and immediately transfers data to a specified I/O slot through a mask. The CPU then resumes evaluating the ladder logic. Figures 5-25 and 5-26 show examples of the address format for an IOM.

Latching Instructions

Latches are used to lock in a condition. For example, if an input contact is on for only a short time, the output coil is on for the same short time. If it were desired to keep the output on even if the input goes low, a latch could be used. This can be done by using the output coil to latch itself on (see Figure 5-27). It can also be done with a special coil called a *latching coil* (see Figure 5-28). When a latching output is used, it stays on until it is unlatched. Unlatching is done with a special coil called an *unlatching coil*. When it is activated, the latched coil of the same number is unlatched.

Rockwell Automation Output Latch Instruction

The output latch instruction (OTL) is a retentive instruction. If this input is turned on, it will stay on even if its input conditions become false. A retentive output can be turned off only by an unlatch instruction. Figure 5-29 shows an output latch in-

Figure 5–24   Output energize instruction (OTE).

O:5

—( )—
     1

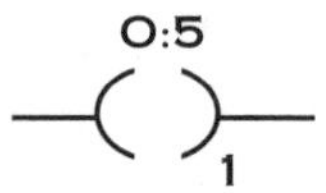

Figure 5–25 Format for an immediate output instruction (IOM) address.

Figure 5–26 Immediate output instruction I/O numbering address.

O:2	Outputs of slot 2, word 0
O:2.1	Outputs of slot 2, word 1
O:1	Outputs of slot 1, word 0

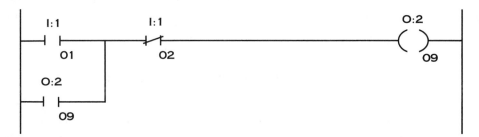

Figure 5–27 Example of a Rockwell Automation latching an output on. If input I:1/01 is true, coil O:2/09 is energized. (Remember that contact I:1/02 is normally closed.) When coil O:2/09 energizes, it latches itself on by providing a parallel path around I:1/01. The only way to turn the latched coil off is to energize normally closed contact I:1/02, which opens the rung and deenergizes coil O:2/09.

Figure 5–28 Example of a Rockwell Automation latch and unlatch instruction. If input I:1/01 is true, output O:2/01 energizes and stays energized even if input I:1/01 becomes false. O:2/01 remains energized until input I:1/02 becomes true and energizes the unlatch instruction (coil O:2/01). Note that the coil number of the latch is the same as that of the unlatch.

Figure 5–29 Output latch instruction (OTL).

Figure 5–30 Output unlatch instruction (OTU).

struction (OTL). In this case if the rung conditions for this output coil are true, the output bit is set to a 1 and remains a 1 even if the rung becomes false. The output is latched on. Note that if the OTL is retentive, a processor power loss turns off the actual output, but when power is restored, the output is retentive and turns on. This is also true of switching from run to program mode. The actual output turns off, but the bit state of 1 is retained in memory. When the processor is switched to run again, retentive outputs turn on again regardless of the rung conditions. Retentive instructions can help or hinder the programmer. Be very careful from a safety standpoint when using retentive instructions. The programmer must use an unlatch instruction to turn a retentive output off.

Rockwell Automation Output Unlatch Instruction

The output unlatch instruction (OTU) is used to unlatch (change the state of) retentive output instructions. It is the only way to turn an output latch instruction (OTL) off. Figure 5-30 shows an output unlatch instruction (OTU). If this instruction is true, it unlatches the retentive output coil of the same number.

PROGRAM FLOW CONTROL INSTRUCTIONS

Many types of program flow control instructions are available on PLCs. Flow control instructions can be used to control the sequence in which the program is executed. They allow the programmer to change the order in which the CPU scans the ladder diagram. These instructions are typically used to minimize scan time and create a more efficient program. They also can be used to help troubleshoot ladder logic. The Programmer should use flow control instructions with great care. Serious consequences can occur if they are improperly used because their use causes portions of the ladder logic to be skipped.

Rockwell Automation Jump and Label Instructions

Rockwell Automation has jump (JMP) and label (LBL) instructions available (see Figure 5-31). They can be used to reduce program scan time by omitting a program section until it is needed. It is possible to jump forward and backward in the ladder, but the programmer must be careful not to jump backward an excessive number of times. A counter, timer, logic, or the program scan register should be used to limit the amount of time spent looping inside a JMP/LBL instruction.

Figure 5–31 Rockwell Automation jump (JMP) and label (LBL) instructions.

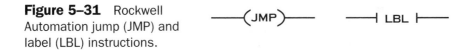

If the rung containing the JMP instruction is true, the CPU skips to the rung containing the specified label (LBL) and continues execution. It is possible to jump to the same label from one or more JMP instructions.

Rockwell Automation Jump-to-Subroutine Instructions   Rockwell Automation also has subroutine instructions available. The jump-to-subroutine (JSR), subroutine (SBR), and return (RET) are used for this purpose (see Figures 5-32 and 5-33). Subroutines can be used to store recurring sections of logic that must be executed in several points in the program. A subroutine saves effort and memory because it is programmed only once. Subroutines can also be used for time-critical logic by using immediate I/O in them.

The SBR instruction must be the first instruction on the first rung in the program file that contains the subroutine. Subroutines can be nested up to 8 deep. This allows the programmer to direct program flow from the main program to a subroutine and then on to another subroutine, and so on. The file number entered in the JSR instruction identifies the desired subroutine. This instruction serves as the label or identifier for a program file as a regular subroutine file. The SBR instruction is always evaluated as true. The RET instruction shows the end of the subroutine file. It causes the CPU to return to the instruction following the previous JSR instruction.

Subroutines can be very useful in breaking a control program into smaller logical sections. For example, the main file we use for our ladder logic is file 2. Many times all of the ladder logic is programmed in file 2. It makes more sense and is more understandable if the program is divided into logical modes of operation. For example, the programmer could make file 2 contain the logic needed for overall system operation: The main ladder logic file (file 2) could contain logic to determine whether the user wants to go into the automatic or manual mode of operation. It could also call another subroutine that monitors the system for safety. Let's use file 3 for the manual mode of operation, file 4 for the automatic mode of operation, and file 5 for the monitoring logic. We would use conditional logic to determine whether to call the manual subroutine or the automatic routine. If the user hits the manual key, the main file calls the subroutine in file 3. If the user hits

Figure 5–32 Rockwell Automation jump-to-subroutine (JSR) instruction.

JSR
JMP TO SUBROUTINE
SBR FILE NUMBER

Figure 5–33 Rockwell
Automation subroutine (SBR)
and return (RET) instructions.

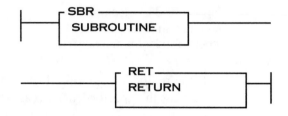

the auto key, the main file calls the automatic subroutine in file 4. We would want
the monitor subroutine to run all of the time. The logic in the main could call the
monitor logic, or the other subroutines can call the monitor mode as subroutines
can be nested and call other subroutines. Think how much simpler it is to program the system when it can be divided into smaller logical sections. It is also
much easier to understand and troubleshoot.

QUESTIONS

1. What is a XIO? XIC?

2. What is an IOM?

3. Write the address for the first input or output for each I/O module.

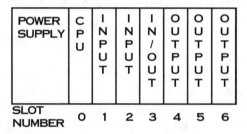

4. Design a latching circuit using a Rockwell Automation latch and unlatch
   instruction. Use contact I:2/7 for the latch input, I:2/8 for the unlatch
   input, and O:5/3 for the coil.

5. Describe the purpose and use of subroutines.

The following are examples of Rockwell Automation SLC addresses. Thoroughly explain each address.

6. B3:16/12

7. O0:1/6

8. I1:1/3

9. O0:2/5

10. I1:1/2
11. T4:7.PRE
12. C5:0.ACC
13. F8:0
14. I1:1/2
15. T4:5.EN
16. T4:3.DN
17. O0:4/9
18. I1:2/4
19. N10:3
20. I:3.0/4
21. I:2/17
22. O:7/12
23. B3:3/5
24. S:42
25. N7:12

PROGRAMMING PRACTICE

These exercises utilize the LogixPro software included with this book. Go to the LogixPro Student Exercises and Documentation. The exercises at the end of Chapter 4 described how to install and utilize the LogixPro Software.

1. Enter the circuit that you designed in question 4 and test it in the I/O simulator (Make sure you use appropriate addresses).

2. Follow the instructions and program and test the Door Simulation— Applying Relay Logic exercise. Document your program.

3. Complete student programming exercise 2 for the Door Simulation exercise. Document your program.

4. Complete student programming exercise 3 for the Door Simulation exercise. Document your program.

EXTRA CREDIT

Complete student programming exercise 4 for the Door Simulation exercise. Document your program.

chapter

Timers and Counters 6

Timers and counters are invaluable in PLC programming. Industry has needs for counting product, controlling time sequences, and so on. This chapter examines the types and programming of timers and counters. Timers and counters are similar in all PLCs.

OBJECTIVES

Upon completion of this chapter you will be able to:
1. Describe the use of timers and counters in ladder logic.
2. Describe at least two types of counters.
3. Define such terms as *retentive, cascade, delay-on,* and *delay-off.*
4. Utilize timers and counters to develop applications.
5. Describe at least two types of timers.

TIMERS

Timing functions are very important in PLC applications. Cycle times are critical in many processes. Timers are used to delay actions and to keep an output on for a specified time after an input has turned it off or to keep an output off for a specified time before turning it on.

Think of a garage light. A nice feature would be a touch and release on the on switch causing the light to turn on immediately and stay on for a given time (maybe 2 minutes). At the end of the time, the light turns off. This allows the person to get into the house while the light is on. In this example, the output (light) turned on instantly when the input (switch) turned on. The timer counted down the time (timed out) and turned the output (light) off. This is an example of a

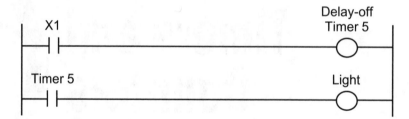

Figure 6–1 Delay-off timing circuit. If contact X1 closes, the delay-off timer immediately turns on, which turns on the light (coil). When the timer reaches the programmed time, it turns off, turning off the light also.

delay-off timer that turns on instantly, counts down, and then turns off. Consider Figure 6-1. When switch X1 is activated, the timer turns on and starts counting time intervals. The timer is used in the second rung as a contact. If the timer is on, the contact in rung 2 (timer 5) closes and turns on the light. When the timer times out (in this case, in 2 minutes), the contact in rung 2 opens and the light turns off. As this example illustrates, output coils can be used as contacts to control other outputs.

The other type is called *delay-on timer*. When input X3 to the delay-on timer is activated, the timer starts timing but remains off until the time has elapsed (see Figure 6-2). In this case, the switch is activated and the timer starts to time, but it remains off until the total time has elapsed. Then the timer turns on, closing the contact in rung 2 and turning on the light.

Many PLCs use *block-style timers and counters* (see Figure 6-3). The timers of most PC brands have been programmed in similar ways. Each timer has a number to identify it. For some the timer is as simple as T7 (timer 7) in Figure 6-3. This chapter utilizes generic examples to illustrate some concepts about

Figure 6–2 Delay-on timing circuit. If contact X3 closes, the timer begins timing. When the time reaches the programmed time, the timer turns on, which turns on the light.

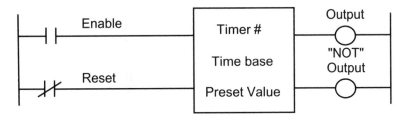

Figure 6–3 Typical block timer.

timers and counters but mainly focuses on Rockwell Automation timers and counters.

Every timer has a *time base*. A timer can typically be programmed with several different time bases: 1.0 second, 0.1 second, and 0.01 second are typical. A programmer who entered 0.1 for the time base and 50 for the number of delay increments, has a timer with a 5-second delay (50×0.1 second = 5 seconds).

Timers also must have a preset value. The *preset value* is the number of time increments the timer must count before changing the state of the output. The actual time delay equals the preset value multiplied by the time base. Presets can be a constant value or a variable. If it is variable, the timer uses the real-time value of the variable to calculate the delay. This allows delays to be changed depending on conditions during operation. An example is a system that produces two different products, each requiring a different time in the actual process. Product A requires a 10-second process time, so the ladder logic assigns 10 to the variable. For product B, the ladder logic can change the value to that required by B. When a variable time is required, a variable number is entered into the timer block. Ladder logic can then assign values to the variable.

Timers typically have one or two inputs. Every timer has one input that functions as a *timer enable input*. When this input is true (high), the timer begins timing. Some timers have a second input used to reset the timer's accumulated time to zero. If the reset line changes state, the timer clears the accumulated value. For example, the timer in Figure 6-3 requires a high to be active. If the reset line goes low, the timer clears the accumulated time to zero. Some timers have only the enable input and utilize a separate reset instruction to reset the accumulated time to zero.

Timers can be retentive or nonretentive. *Retentive timers*, sometimes called *accumulating timers*, do not lose the accumulated time when the enable input line goes low but retain it until the line goes high again. When it does, they add to the count. *Nonretentive timers* lose the accumulated time when the enable input goes low. When this happens, the timer count goes to zero. Retentive timers function like a stopwatch, which can be started and stopped and retain its timed value. A stopwatch also has a reset button to reset the time to zero.

Figure 6–4   Numbering
system format for a timer.

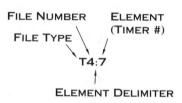

Rockwell Automation PLC-5, SLC 500, and MicroLogix 1000 Timers

Timer On-Delay Instruction   The *timer on-delay* instruction turns an output on after a timer has been on for a preset time interval. The timer on-delay (TON) begins accumulating time when the rung becomes true and continues until one of the following conditions is met: The accumulated value equals the preset value, the rung goes false, or a reset timer instruction resets the timer. Figure 6-4 shows the numbering system for timers. *T* stands for timer, *4* is the file number of the timer, and *7* is the actual timer number.

Status Bit Use   Ladder logic can use timer status bits. Several bits are available (see Figure 6-5). Consider timer T4:0. The *timer enable bit* (.EN) is set immediately when the rung goes true and stays set until the rung goes false or a reset instruction resets the timer. The .EN bit indicates that the timer is enabled. The .EN bit from any timer can be used for logic. For example, T4:0.EN can be used as a contact in a ladder.

The *timer timing bit* (.TT) can also be used. It is set when the rung goes true and remains true until the rung goes false or the .DN bit is set (accumulated value equals preset value). For example, T4:0.TT can be used as a contact in a ladder. The *timer done bit* (.DN) is set until the accumulated value equals the

Condition	Result
If the rung is true	.EN bit remains set (1) .TT bit remains set (1) .ACC value is cleared and begins counting up
If the rung is false	.EN bit is reset .TT bit is reset .DN bit is reset .ACC value is cleared and begins counting up

Figure 6–5   Conditions of special timer bits.

Time Base	Potential Time Range
1.0 Seconds	To 32,767 time-base intervals (up to 9.1 hours)
.01 Second (10 ms)	To 32,767 time-base intervals (up to 5.5 minutes)

Figure 6–6   Time-base intervals available.

preset value and the rung goes false when a reset instruction resets the timer. When the .DN bit is set, the timing operation is complete. For example, T4:0.DN could be used as a contact in the ladder, and the preset (.PRE) is also available. For example, T4:0.PRE accesses the preset value of T4:0. Note that .PRE value is an integer.

Accumulated Value Use   The programmer can also use the accumulated value (.ACC). It is acquired in the same manner as the status bits and preset. For example, T4:0.ACC accesses the accumulated value of timer T4:0.

Time Bases   Two time bases are available: intervals of 1.0 second and 0.01 second (see Figure 6-6; the potential time ranges are also shown). If a longer time is needed, timers can be cascaded (discussed later in this chapter). Figure 6-7 shows how timers are handled in memory. This timer uses three bits in the first storage location to store the present status of the timer bits (.EN, .TT, and .DN). The preset value (.PRE) is stored in the second 16 bits of this timer storage. The third 16 bits hold the accumulated value of the timer.

Figure 6-8 shows a TON timer used in a ladder logic program. When input I:2/3 is true, the timer begins to increment the accumulated value of TON timer 4:0 in 1-second intervals. The timer timing bit (.TT) for timer 4:0 is used in the second rung to turn on output O:5/1 while the timer is timing (.ACC < .PRE). The

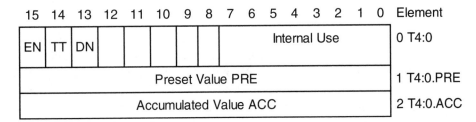

Figure 6–7   Use of control words for timers.

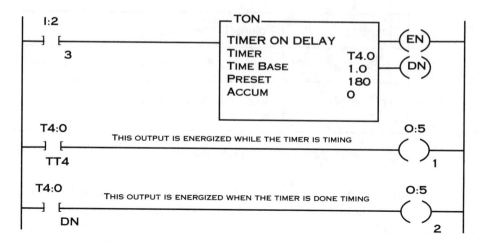

Figure 6–8 TON timer used in a ladder logic program.

timer done bit (.DN) of timer 4:0 is used in rung 3 to turn on output O:5/2 when the timer has finished timing (.ACC = .PRE). The preset for this timer is 180, which means that the timer must accumulate 180 total 1-second intervals to time out. Note that this is not a retentive timer. If input I:2/3 goes low before 180 is reached, the accumulated value is reset to zero.

Timer Off-Delay Instruction   The *timer off-delay (TOF)* instruction turns an output on or off after the rung has been off for a desired time. The TOF instruction starts to accumulate time when the rung makes a true-to-false transition and continues until the accumulated value equals the preset value, the rung becomes true, a reset timer instruction resets the timer. The timer enable bit (.EN bit 15) is set when the rung becomes true and is reset when the rung becomes false, a reset instruction resets the timer. The timer timing bit (.TT bit 14) is set when the rung becomes false and .ACC < .PRE. The .TT bit is reset when the rung go true, or the .DN bit is reset (.ACC = .PRE).

The done bit (.DN bit 13) is set when rung conditions are true. It remains set until rung conditions go false and the accumulated value is greater than or equal to the preset value.

Figure 6-9 shows the use of a TOF timer in a ladder logic diagram. Input I:2/3 enables the timer. When input I:2/3 makes a true-to-false transition, the accumulated value is incremented so long as the input stays false and .ACC <= .PRE. The timer timing bit for timer T4:0 (T:4.TT) turns on output O:5/1 while the accumulated value is less than the preset value. The done bit (.DN) for timer 4:0 (T4:0.DN) turns on output O:5/2 when the timer has completed the timing (.ACC = .PRE).

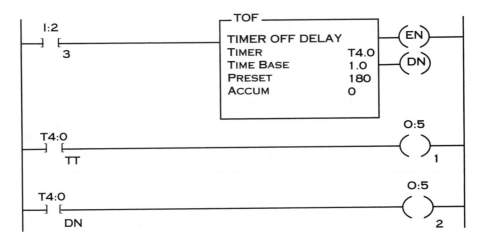

Figure 6–9 TOF timer used in a ladder logic diagram.

Retentive Timer On Instruction

The *retentive timer on (RTO)* instruction turns an output on after a set time period (see Figure 6-10). The RTO timer is an accumulating timer; it retains the accumulated value even if the rung goes false, power is lost, modes are switched or the associated SFC becomes inactive. The only way to zero the accumulated value is to use a reset instruction in another rung with the same number as the RTO to be reset.

The status bits can be used as contacts in a ladder diagram. The timer enable bit (.EN) is set when the rung becomes true. When the .EN bit is a 1, it indicates

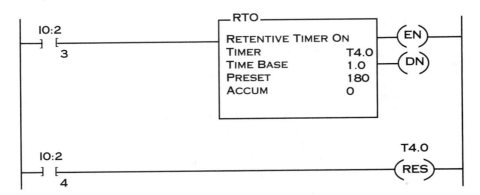

Figure 6–10 Use of an RTO timer.

that the timer is timing. It remains set until the rung becomes false or a reset instruction zeros the accumulated value.

The timer timing bit (.TT) is set when the rung becomes true and remains set
until the accumulated value equals the preset value or a reset instruction resets the
timer. When the .TT bit is a 1, it indicates that the timer is timing.

The .TT bit is reset when the rung becomes false or when the done bit (.DN) is
set. The timer done bit (.DN) is set when the timer's accumulated value equals the
preset value. When the .DN bit is set, the timing is complete. The .DN is reset
with the reset instruction.

Cascading Timers

Applications sometimes require longer time delays than one timer can accomplish. Multiple timers then can be used to achieve a longer delay than otherwise
possible. One timer acts as the input to another. When the first timer times out, it
becomes the input to start the second timer timing. This is called *cascading*.
Figure 6-11 shows a generic example of cascading timers.

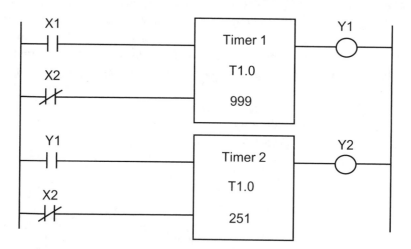

Figure 6–11   Cascading accomplished by using two timers to extend
the time delay. The first timer output, Y1, acts as the input to start the
second timer. When input X1 becomes true, timer 1 begins to count to 999
seconds. (The limit for this particular brand of timers is 999 seconds and
the accumulated time for accumulating timers is 99,999,999.) When it
reaches 999, output Y1 turns on, activating input Y1 to timer 2. Timer 2 then
counts to 251 seconds and then turns on output Y2. The delay was 1250
seconds.

COUNTERS

Counting functions are very important in industrial applications. Often a product action must be counted so that another action can take place. For example, if 24 cans go into a case, the PLC should sense the 24th can and seal the case. Almost all applications require counters.

Several types of counters are available, including up counters, down counters, and up/down counters. The choice of which to use depends on the task to be done. For example, to count the finished product leaving a machine, we might use an *up counter*. To track how many parts are left, we might use a *down counter*. When using a PLC to monitor an automated storage system, we might use an *up/down counter* to track how many parts are coming and how many are leaving to establish the actual total number in stock.

Counters normally use a low-to-high transition from an input to trigger the counting action. Figure 6-12 shows a generic up-counter example. Counters have a reset input or a separate reset instruction to clear the accumulated count. Counters are very similar to timers: Counters count the number of low-to-high transitions on the input line and timers count the number of time increments.

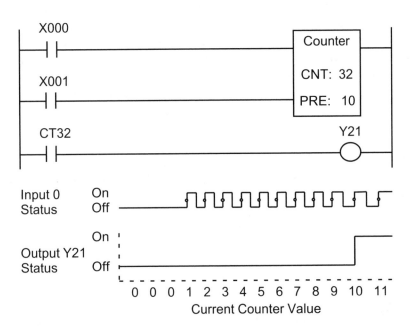

Figure 6–12 Working a generic up-counter. When 10 or more low-to-high transitions of input X000 have been made, counter CT32 is energized, which energizes output Y21.

Note that the counter in Figure 6-12 is *edge-sensitive triggered*. The rising, or leading edge, triggers the counter. X000 is used to count the pulses. Every time an off-to-on transition on X000 occurs, the counter adds 1 to its count. When the accumulated count equals the preset value, the counter turns on, which turns on output Y21. X001 is used as a reset/enable. If contact X001 is closed, the counter returns to zero. The counter is active (enabled or ready to count) only if X000 is off (open).

Some brands of PLCs also have an up/down counter, which causes a count to decrease by one every time a pulse occurs. An up/down counter has one input that causes it to increment the count and another that causes it to decrement the count.

Some ladder diagram statements can utilize these counts for comparing AND/OR decision making as well as comparing constants or variables to control outputs.

Allen-Bradley PLC-5, SLC 500, and MicroLogix 1000 Counters

A Rockwell Automation counter has a timer number, a preset, and an accumulated value. The counter is numbered like the timer except that it begins with a *C*. The next number is a file number. The default file number for counters is *5*. The third value is the counter number, in this case *4* (see Figure 6-13).

A ladder diagram can access counter status bits, presets, and accumulated values (see Figure 6-14). It can use the .CU, .CD, .DN, .OV, or .UN bit for logic. It can also use the preset (.PRE) and the accumulated count (.ACC).

Counter values are stored in three 16-bit words of memory. The first 8 bits of the first word are only for the CPU's internal use. The most significant bits of the first word store the status of certain bits associated with the counter (see Figure 6-15). The count-up enable bit (.CU bit 15) indicates that the counter is enabled and is reset when the rung becomes false or when it is reset by a RES instruction. When high, the count-up done bit (.DN bit 13) indicates that the accumulated count has reached the preset value. It remains set even when the accumulated value (.ACC) exceeds the preset value (.PRE). The .DN bit is reset by a reset (RES) instruction.

The CPU sets the count-up overflow bit (.OV bit 12) to show that the count has exceeded the upper limit of +32,767. When this happens, the counter's accumulated value "wraps around" to −32,768 and begins to count up from there. The .OV bit can be reset with a reset (RES) instruction.

Figure 6-16 shows the use of a *count-up counter (CTU)* in a ladder diagram. As each time input I:2/3 makes a low-to-high transition, the counter's accumulated value is incremented by 1. The done bit of counter 5:0 (C5:0.DN) turns output O:5/1 on when the accumulated value equals the preset value (.ACC = .PRE).

Figure 6–13 Addresses of counters.

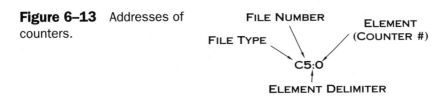

Figure 6–14 Examples of uses for counter values.

C5:4.DN   Use of the Done Bit
C5:4.PRE   Use of the Preset Value
C5:4.ACC   Use of the Accumulated Value

Figure 6–15 Counter values and status bits are stored in memory.

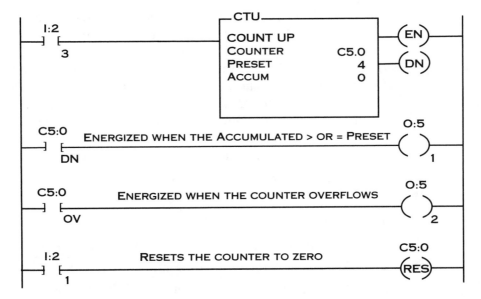

Figure 6–16 Use of a count-up counter (CTU) in a ladder diagram.

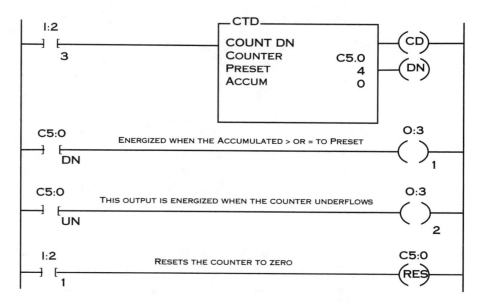

Figure 6–17   Use of a count-down counter (CTD) in a ladder diagram.

The overflow bit of counter 5:0 (C5:0.OV) sets output O:5/2 to on if the count ever reaches +32,767. The last rung uses input I:2/1 to reset counter 5's accumulated value to zero.

Figure 6-17 shows the use of a *count-down counter (CTD)* in a ladder diagram. Each time input I:2/3 makes a low-to-high transition, the counter's accumulated value is decremented by 1. The done bit of counter 5:0 (C5:0.DN) turns on output O:3/1 when the accumulated value equals or exceeds the preset value (.ACC = .PRE). The accumulated value of a counter is retentive and is retained until a reset instruction is used. The underflow bit of counter 5:0 (C5:0.UN) sets output O:3/2 to on if the count ever underflows −32,768. Note the use of the reset instruction to reset the accumulated value of the counter to zero. The last rung uses input I:2/1 to reset counter 5's accumulated value to zero (see Figure 6-18).

QUESTIONS

1. For what are timers typically used?
2. Explain the two types of timers and how each might be used.
3. What does the term *retentive* mean?
4. Draw a typical retentive timer and describe the purpose of the inputs.
5. For what are counters typically used?

Reset Instruction Use	The CPU Resets
Timer (Do not use a reset instruction for a TOF timer.)	.ACC Value .EN Bit .TT Bit .DN Bit
Counter	.ACC Value .EN Bit .OV or .UN Bit .DN Bit

Figure 6–18   Effect of the use of a reset (RES) instruction on timer and counter values and bits.

6. In what way(s) are counters and timers very similar?

7. Explain the two contacts usually required for a counter.

8. What is *cascading*?

9. You have been asked to program a system that requires completed parts to be counted. The largest counter available in the PLC's instruction set can count only to 999. Your system must be able to count to 5000. Draw a ladder diagram that shows the method you would use to complete the task. (*Hint:* Use two counters, one as the input to the other. The total count involves looking at the total of the two counters.)

10. Figure 6-19 is a partial drawing of a heat-treat system. You have been asked by your supervisor to troubleshoot the system. The engineer who originally developed the system no longer works for the company and never fully documented it. A short description of the system follows. Study the drawings and data and complete this assignment.

    A part enters this portion of the process. The temperature must then be raised from room temperature to 500 degrees Fahrenheit. There is also a part presence sensor and another sensor that turns on when the temperature reaches 500 degrees. The part must then be pushed out of the machine. The cycle should take about 25 seconds. If an excessive amount of time is required but the temperature has still not been reached, an operator must be informed and must reset the system. Study the system drawing, I/O chart (Figure 6-20), and ladder diagram (Figure 6-21), and then complete the I/O table and answer the questions.

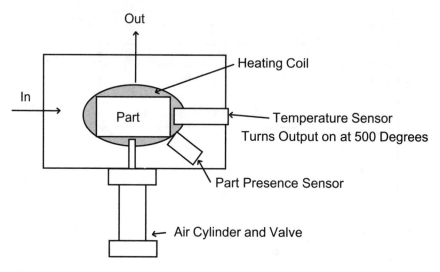

Figure 6–19   Partial drawing of a heat-treat system.

Figure 6–20   I/O table.

System I/O	
X3	Part Presence Sensor
X20	Operator Reset
Y5	
Y8	
Timer 1	
Timer 9	
Counter 1	
Counter 2	

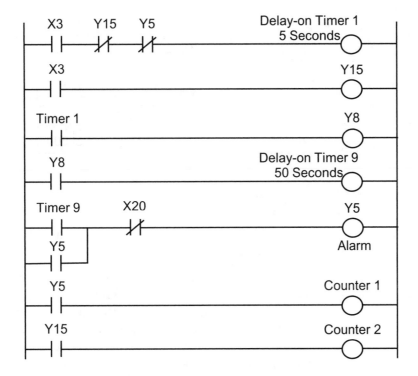

Figure 6–21 Ladder diagram.

Complete the I/O chart in Figure 6-20 by writing short comments that describe the purpose of each input, output, timer, and counter. Make your comments very clear and descriptive so that the next person to troubleshoot the system will have an easier task. Then refer to Figure 6-21 and answer the following questions.

   a. What is the purpose of counter 1?
   b. What is the purpose of input X3?
   c. What is input X20 being used for?
   d. What is the purpose of counter 2?
   e. What is the purpose of contact Y15?
   f. Part of the logic is redundant. Identify that part and suggest a change.

11. Examine the ladder diagram in Figure 6-22. Assume that input 00001 is always true. What will this ladder logic do?

12. You have been assigned to develop a stoplight application. Your company thinks there is a large market in intelligent street corner control and is developing a PLC-based system that will adapt the timing of lights to

```
                                                    Timer 0
    X1      Timer 5                                 1 Second
   | |        |/|                                      O

            Timer 0                                 Timer 5
                                                    5 Seconds
            | |                                        O

            Timer 0                                    Y11
            | |                                        O
```

Figure 6–22 Ladder diagram.

traffic volume. Your task is to program a normal stoplight sequence to use to compare against the new system.

Note that Figure 6-23 really has two sets of lights. The north–south lights must react exactly alike, and the east–west set must complement the north–south set. Write a program that will keep the green light on for 25 seconds, the yellow on for 5 seconds, and the red on for 30 seconds. You must also add a counter because the bulbs are replaced at a certain count for preventive maintenance. The counter should count complete cycles. (*Hint*: To simplify your task, do one small task at a time. Do not try to write the entire application at once. Write ladder logic to get one light working, then the next, and then the next before you even worry about the other stoplight. When you finish one set, the other is a snap. *Remember:* A well-planned job is half done.)

Figure 6–23 Stoplight application.

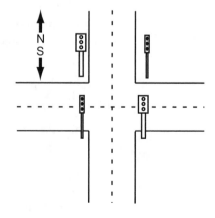

Write the ladder diagram. Thoroughly document the ladder with labels and rung comments.

PROGRAMMING PRACTICE

These exercises utilize the LogixPro software included with this book. Go to the LogixPro Student Exercises and Documentation. Remember that the Rockwell Automation instruction help is also available.

1. This exercise involves PLC Timers—Introductory Exercise.
2. Complete the Traffic Control exercise.
3. Complete the PLC Counters—Introductory Exercise.

chapter

Input/Output Modules and Wiring

7

Originally, PLCs were designed strictly for simple digital (on/off) control. Over the years, manufacturers have added to the PLC's capabilities. Today, I/O cards are available for almost any application imaginable.

OBJECTIVES

Upon completion of this chapter, you will be able to:
1. Define such terms as *resolution, high density, discrete,* and *RF*.
2. Choose an appropriate I/O module for a given application.
3. Describe the difference between a sinking and sourcing module.

Originally, PLCs were used to control one simple machine or process. The changes in U.S. manufacturing have required much more capability. In addition, the increasing speed of production and the demand for higher quality require closer control of industrial processes. PLC manufacturers have added modules to meet these new control requirements. Special modules have been developed to meet almost any imaginable need, such as temperature control.

Industry is beginning to integrate its equipment to share data. Modules have been developed to allow the PLC to communicate with other devices, such as computers, robots, and machines.

Velocity and position control modules have been developed to meet the needs of accurate high-speed machining. These modules also make it possible for entrepreneurs to start new businesses that design and produce special-purpose manufacturing devices, such as packaging equipment, palletizing equipment, and various other production machinery. These modules are also designed to be easy to use. They are intended to make it easier for the engineer to build an application. The balance of this chapter examines many of the modules that are available.

DIGITAL (DISCRETE) MODULES

Digital modules are also called *discrete modules* because they are either on or off. A large percentage of manufacturing control can be accomplished through on/off control. Discrete control is easy and inexpensive to implement.

Digital Input Modules

These modules accept an "on" or "off" state from the real world. The input modules are attached to devices such as switches or digital sensors. The modules must be able to buffer the CPU from the real world. Assume that the input is 250 volt AC. The input module must change the 250 volt AC level to a low-level DC logic level for the CPU. The modules must also optically isolate the real world from the CPU. Input modules usually have fuses for module protection.

Input modules typically have light-emitting diodes (LEDs) for monitoring the inputs, with one LED for every input. If the input is on, the LED is on. Some modules also have fault indicators. The fault LED turns on if the module has a problem. The LEDs on the modules are very useful for troubleshooting.

Most modules also have plug-on wiring terminal strips to which all wiring is connected. The terminal strip is plugged onto the actual module. If a module has a problem, the entire strip is removed, a new module inserted, and the terminal strip plugged into the new module; no rewiring is needed. A module can be changed in minutes or less. This is vital considering the huge cost of a system being down. (The term *down* means "unable to produce product.")

With the exception of small PLCs, input modules usually need to be supplied with power. The power must be supplied to a common terminal on the module through an input device and back to a specific input on the module. Current must enter at one terminal of the I/O module and exit at another terminal (see Figure 7-1). The figure shows a power supply, field input device, the main path, I/O circuit, and the return path. Unfortunately, this would require two terminals for every I/O point, but most I/O modules provide groups of I/O that share return paths (commons). Figure 7-2 shows the use of a common return path.

Some modules provide multiple commons. This allows the user to use two voltages on the same module (see Figure 7-3). These commons can be jumpered together if desired (see Figure 7-4).

When load-powered sensors are used, a small leakage current is always necessary for the operation of the sensor. This is not normally a problem. In some cases, however, this leakage current is enough to trigger the input of the PLC module. In this case a resistor can be added to "bleed" the leakage current to ground (see Figure 7-5). When a bleeder resistor is added, most of the current

Figure 7–1 Typical I/O current path. (*Courtesy PLC Direct by Koyo.*)

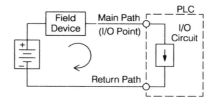

Figure 7–2 Typical shared return path (common). (*Courtesy PLC Direct by Koyo.*)

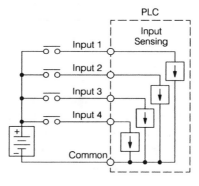

Figure 7–3 Input module with dual commons that allow the user to mix input voltages.

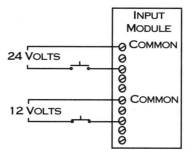

Figure 7–4 Dual commons can be wired together so that all inputs use the same voltage.

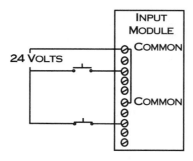

Figure 7–5 A bleeder
resistor.

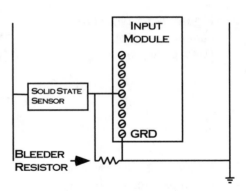

goes through it to common. This ensures that the PLC input turns on only when
the sensor is really on.

Figure 7-6 shows an input module for a Rockwell Automation SLC. Note that
the commons are connected internally for this module and that the negative side
of the DC voltage must be connected to the common. The positive voltage from
the power supply is brought through the input devices and back to the input terminal. This is a *sinking module.*

Digital Output Modules

Discrete output modules are used to turn real-world output devices on or off. Discrete
output modules, which can be used to control any two-state device, are available in
AC and DC versions as well as various voltage ranges and current capabilities.

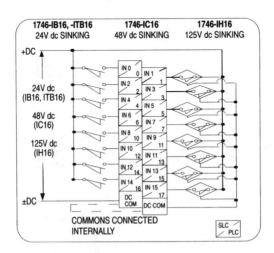

Figure 7–6 SLC DC input module wiring. (*Courtesy Rockwell Automation Inc.*)

The current specifications for a module are normally given as an overall module current and as individual output current. The specification may rate each output at 1 amp. If it is an eight-output module, we might assume that the overall current limit is 8 amp (8×1), but this is normally not the case. The overall current limit is probably less than the sum of the individuals. For example, the overall current limit for the module might be 5 amp. The user must be careful that the total current to be demanded does not exceed the total that the module can handle. Normally, none of the user's outputs draws the maximum current, nor are all normally on at the same time. The user should consider the worst case when choosing an appropriate output module.

Output modules are normally fused to provide short circuit protection for wiring only to external loads. If there is a short circuit on an output channel, the output transistor, triac, or relay associated with that channel will likely be damaged. In that case the module must be replaced or the output moved to a spare channel on the output module. Fuses are normally easily replaced; the technical manual for the PLC explains the exact procedure. Figure 7-7 shows the fuse location and jumper setting for a Rockwell SLC module. Note that by choosing the jumper location, the user can decide whether the processor faults or continues in the event that a fuse blows.

Some output modules provide more than one common terminal. This allows the user to use different voltage ranges on the same card (see Figure 7-8). These multiple common terminals can be tied together if the user desires, but all outputs must use the same voltage (see Figure 7-9).

Output modules can be purchased with transistor output, triac output, or relay output. The transistor output is used for DC outputs. Various voltage ranges and current ranges are available, as well as transistor-transistor logic (TTL) output. Triac outputs are used for AC devices and are available in various voltage ranges and current ranges. The relay outputs are found quite often on small PLCs and are available for large ones. Relay outputs can be used with AC or DC voltages. The voltages can even be mixed (see Figure 7-10). Figure 7-11 shows the use of a bleeder resistor to "bleed off" unwanted leakage through an output module. This leakage can occur with the solid-state devices used in output modules. Figure 7-12 illustrates the inside of the relay module; note the relays. Relay outputs should be protected with a diode to prolong contact life (see Figure 7-13).

See Figure 7-14 for a DC output module for a Rockwell Automation SLC. Note that the positive side of the DC voltage must be connected to the VDC terminal. The output devices are connected to an output terminal and then to the negative side of the power supply. The negative side of the supply must also be connected to the DC common.

Figure 7-15 is an AC output module for a Rockwell Automation SLC. Note that L1 is connected to VAC 1 AND VAC 2. Outputs are connected to output terminals and then to L2.

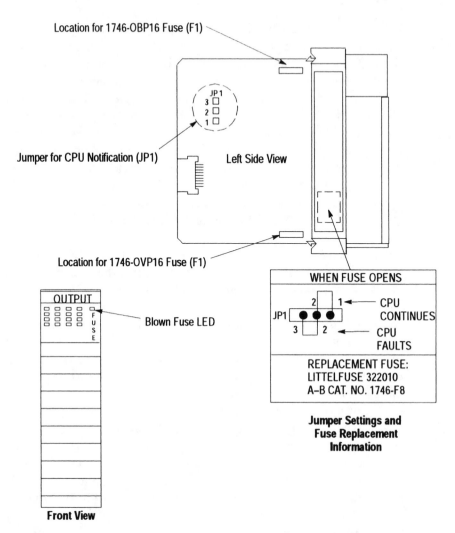

Figure 7–7 Fuse and jumper settings for a Rockwell SLC. The technical manual for each PLC explains the exact procedure. (*Courtesy Rockwell Automation Inc.*)

Figure 7–8 Use of a dual-common output module.

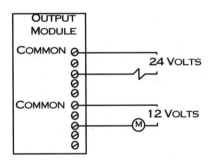

Figure 7–9 Dual-common output module with the commons tied together which requires the voltages to be the same.

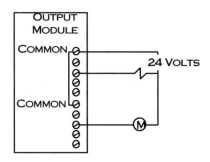

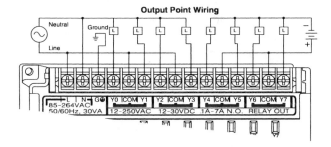

Figure 7–10 Wiring of a relay output module. (*Courtesy of PLC Direct by Koyo.*)

Figure 7–11 Bleeder resistor "bleeds off" unwanted leakage through an output module.

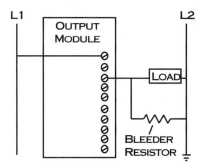

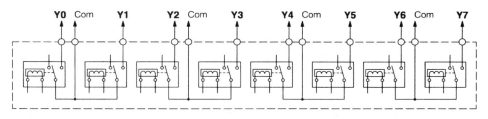

Figure 7–12 Wiring of relay outputs. (*Courtesy PLC Direct by Koyo.*)

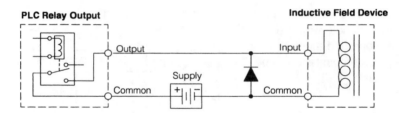

Figure 7–13   A diode used to help prolong contact life. (*Courtesy PLC Direct by Koyo.*)

Some applications require connecting a PLC output to the solid state input of a device. This is usually done to provide a low-level signal, not to power an actuator or coil. Figure 7-16 shows the connection of a solid state sourcing input to a PLC sinking output on an output device. Figure 7-17 shows a sinking output connected to a solid state input on an output device. It is important to size the pull-up resistor properly. Figure 7-18 shows the formula for calculating the proper size of pull-up resistor.

High-Density I/O Modules

High-density modules are digital I/O modules. A normal I/O module has eight inputs or outputs; a high-density module may have up to 32 inputs or outputs. The advantage is that a PLC rack has a limited number of slots. Each module uses a

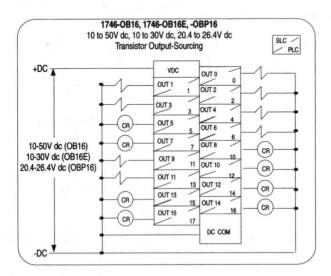

Figure 7–14   Example of SLC DC output module wiring. (*Courtesy Rockwell Automation Inc.*)

Figure 7–15 SLC AC output module wiring with triac outputs. (*Courtesy Rockwell Automation Inc.*)

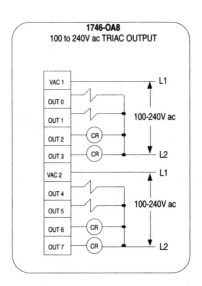

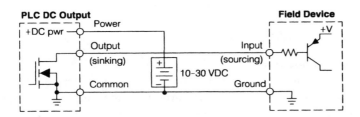

Figure 7–16 PLC sinking output connected to a solid state sourcing input on an output device. (*Courtesy PLC Direct by Koyo.*)

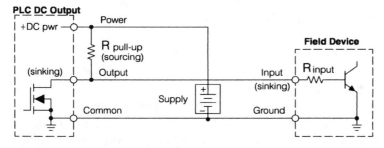

Figure 7–17 PLC sinking output connected to a solid state sinking input on an output device. (*Courtesy PLC Direct by Koyo.*)

$$I_{input} = \frac{V_{input\ (turn-on)}}{R_{input}}$$

$$R_{pull-up} = \frac{V_{supply} - 0.7}{I_{input}} - R_{input} \qquad\qquad P_{pull-up} = \frac{V_{supply}^2}{R_{pullup}}$$

Figure 7–18   Formula for calculating the correct size of resistor. (*Courtesy PLC Direct by Koyo.*)

slot. With the high-density module, it is possible to install 32 inputs or outputs in one slot. The only disadvantage is that the high-density output modules cannot handle as much current per output as normal modules can.

ANALOG MODULES

Computers (PLCs) are digital devices, so they do not work with analog information. Analog data such as temperature must be converted to digital information before the computer can work with it.

Analog Input Modules

Cards called *analog-to-digital (A/D) input cards* have been developed to convert analog information to digital information. Two basic types are available: current sensing and voltage sensing. These cards take the output from analog sensors (such as thermocouples) and change it to digital data for the PLC.

Voltage input modules are available in two types, unipolar and bipolar. *Unipolar modules* can take only one polarity for input. For example, if the application requires the card to measure only 0 to +10 volts, a unidirectional card will work. The *bipolar* card takes input of positive and negative polarity. For example, if the application produces a voltage between −10 and +10 volts, a bidirectional input card is required; because the measured voltage could be negative or positive. Analog input modules are commonly available in 0 to 10 volt models for the unipolar and −10 to +10 volts for the bipolar model.

Analog models are also available to measure current, typically 4 to 20 milliamps. Four milliamps represent the smallest input value, and 20 milliamps represent the largest input value.

The user can configure many analog modules, such as dip switches or jumpers used to configure the module to accommodate different voltages or current. Some manufacturers make modules that will accept voltage or current for input. The user simply wires the module to either the voltage or the current terminals, depending on the application.

Resolution in Analog Modules *Resolution* can be thought of as how closely a quantity can be measured. Imagine a 1-foot ruler. If the only graduations on the ruler are inches, the resolution is 1 inch. If the graduations were every one-fourth inch, the resolution would be one-fourth inch, meaning the closest we can measure any object is one-fourth inch. This is the basis for measuring an analog signal. The computer can work only with digital information. The analog-to-digital (A/D) card changes the analog source into discrete steps.

Examine Figure 7-19(a). Ideally, the PLC is able to read an exact temperature for every setting of the thermostat, but it can work only digitally. Consider Figure 7-19(b). The analog input card changes the analog voltage (temperature) into digital steps. In this example, the analog card changed the temperature from 40 degrees to 160 degrees in four steps. The PLC reads a number between 1 and 4 from the A/D card. A simple math statement in the ladder could change the number into a temperature. For example, assume that the temperature is 120 degrees. The A/D card outputs the number 3. The math statement in the PLC multiplies the number by 40 to get the temperature. In this case, it is 3×40, or 120 degrees. If the PLC reads 4 from the A/D card, the temperature is 4×40, or 160. Assume now that the temperature is 97. The A/D card outputs the number 2, which the PLC reads and multiplies by 40. The PLC believes the temperature to be 80 degrees. The closest the PLC can read the temperature is about 20 degrees if four steps are used. (The temperature that the PLC calculates is always in a range from 20 degrees below

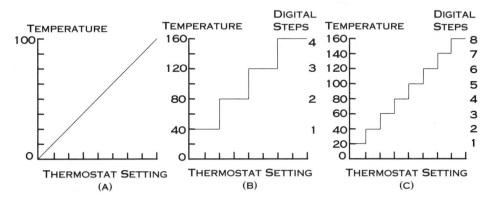

Figure 7–19 Graphs of temperature versus thermostat setting. Note that graph (a) represents a linear relationship of temperature versus setting. In reality, when analog control is used, the analog is really a series of steps (resolution). Graph (b) illustrates a four-step system. The resolution is 40 degrees per step. Graph (c) is an eight-step system whose resolution is 20 degrees.

the actual temperature to 20 degrees above the measured temperature.) The smallest temperature for each step is 40 degrees, which is also the *resolution*.

Consider Figure 7-19(c). This A/D card has eight steps. The resolution is twice as fine as the four-step card, or 20 degrees. For a temperature of 67 degrees, the A/D outputs 3. The PLC multiplies 3×20 and assumes the temperature to be 60. The largest possible error is approximately 10 degrees. (The PLC-calculated temperature is within 10 degrees below the actual temperature to 10 degrees above the actual temperature.) However, industry requires much finer resolution. Typically, an industrial A/D card for a PLC has 12-bit binary resolution, which requires 4,096 steps to measure the analog quantity. Very fine resolution! Cards with even finer resolution are available. The typical A/D card has 12 bits (4,096 steps) or 14 bits (16,384 steps).

Analog modules that can take between one and eight individual analog inputs are available, as are special-purpose A/D modules. One example is the thermocouple module which is just an A/D module adapted to meet the needs of thermocouple input. Thermocouples output very small voltages. To be able to use the entire range of the module resolution, a thermocouple module amplifies the small output from the thermocouple so that the entire 12-bit resolution is used. The modules also provide cold junction compensation. Cold junction compensation means that the module will automatically adjust to changes in ambient temperature so that the actual thermocouple temperature is accurate. These modules are available to enable the acceptance of input from various types of thermocouples. Figure 7-20 shows a tank-filling application.

Analog Output Modules

Analog output modules are also available. The PLC processes digital numbers so it outputs a digital number (STEP) to the digital-to-analog (D/A) converter module. The D/A converts the digital number from the PLC to an analog output. Ana-

Figure 7–20   A tank-filling system.

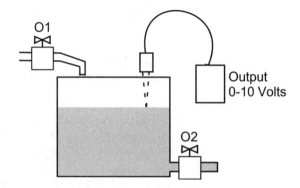

log output modules are available with voltage or current output. Typical outputs are 0 to 10 volts, -10 to $+10$ volts, and 4 to 20 milliamps.

Imagine a bakery. A temperature sensor (analog) in the oven could be connected to an A/D input module in the PLC. The PLC reads the voltage (steps) from the A/D card to learn the temperature. The PLC can then send digital data to the D/A output module, which controls the heating element in the oven. This creates a fully integrated, closed-loop system to control the oven temperature.

REMOTE I/O MODULES

Special modules are available for PLCs that allow the I/O module to be positioned separately from the PLC. In some processes, it is desirable (or necessary) to position the I/O module at a different location, sometimes spread over a wide physical area. In these cases positioning the I/O modules away from the PLC may be desirable. Figure 7-21 shows an example of the use of an Allen-Bradley remote I/O adapter module.

Twisted pair wiring is a common method for connecting the remote module to the processor. Two wires are twisted around each other and connected between the PLC and the remote I/O. Twisting reduces the possibility of electrical interference (noise) because any noise acts on both conductors equally. Twisted-pair connections can transmit data thousands of feet.

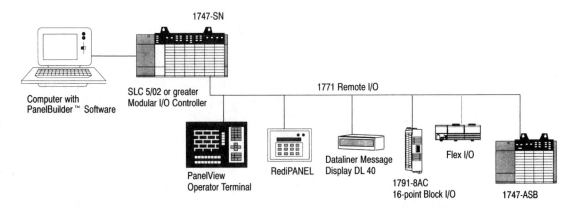

Figure 7–21 Allen-Bradley remote I/O module. (*Courtesy Rockwell Automation Inc.*)

Figure 7–22 Operator display terminal. (*Courtesy Rockwell Automation Inc.*)

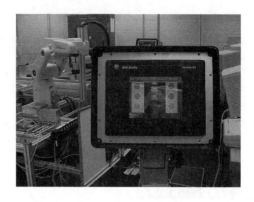

OPERATOR INPUT/OUTPUT DEVICES

As systems become more integrated and automated, they become more complex, and operator information becomes crucial. Many devices provide this information interchange.

Operator Terminals

Many PLC makers now offer their own operator terminals from very simple to very complex. The simple ones are able to display a short message, and the more complex models display graphics and text in color while taking operator input from touch screens (see Figure 7-22), bar codes, keyboards, and so on. These display devices can cost from several hundred to several thousand dollars. If the PLC has most of the valuable information about the processes it controls in its memory, the operator terminal can be a window into the memory of the PLC.

The greatest advances in operator terminals have been in the ease of use. Many PLC manufacturers make available software that runs on an IBM personal computer. The software essentially writes the application for the user, who draws the screens and decides which variables from the PLC to display. The user also decides what input is needed from the operator. After the screens have been designed, they are downloaded to the display terminal.

These smart terminals can store hundreds of pages of displays in their memory. The PLC simply sends a message that tells the terminal which page and information to display. This helps reduce the load on the PLC because the PLC only requests the correct display data, which the display holds in memory, and the terminal displays it. The PLC needs only to update the variables that may appear on the screen. The typical display includes graphics showing a portion of the process, variables showing times or counts, and any other information that might aid an operator in the operation or maintenance of a system.

QUESTIONS

1. What voltages are typically available for I/O modules?
2. If an input module senses an input from a load-powered sensor when it should not, what might be the possible problem, and what could you do about it?
3. List at least three types of output devices available in output modules.
4. Explain the purpose of A/D modules and resolution.
5. Explain the purpose of D/A modules and resolution.
6. What type of module could be used to communicate with a computer?
7. If noise were a problem in an application, what types of changes might help alleviate the problem?
8. What are remote I/O modules?
9. Define the term *resolution*.
10. Why is the use of operator terminals increasing?

chapter

Arithmetic and Advanced Instructions

8

Arithmetic instructions are vital in the programming of systems and can simplify the programmer's task. This chapter focuses on compare, add, subtract, multiply, and divide instructions. It also includes instructions for several PLC brands.

OBJECTIVES

Upon completion of this chapter you will be able to:
1. Describe typical uses for arithmetic instructions.
2. Explain the use of compare instructions.
3. Write ladder logic programs involving arithmetic instructions.
4. Write ladder logic using sequencers.

Contacts, coils, timers, and counters often fall short of what the programmer needs. Many applications require mathematical computation. For example, imagine a furnace application that requires the furnace heat to be between 250 and 255 degrees (see Figure 8-1). If the temperature variable is above 255 degrees, we turn off the heater coil. If the temperature is below 250, we turn on the heater coil. If the temperature is between 250 and 255 degrees, we turn on a green indicator lamp. If the temperature falls below 240 degrees, we sound an alarm. (*Note:* Industrial temperature control is normally more complex than this. Complex process control is covered in Chapters 12 and 13.)

This simple application requires the use of relational operators (arithmetic comparisons) and involves tests of limits (250–255) and the less than function. The use of arithmetic statements makes this application very easy to write. Many of the small PLCs do not have arithmetic instructions available, but all of the larger PLCs offer a wide variety of arithmetic instructions.

127

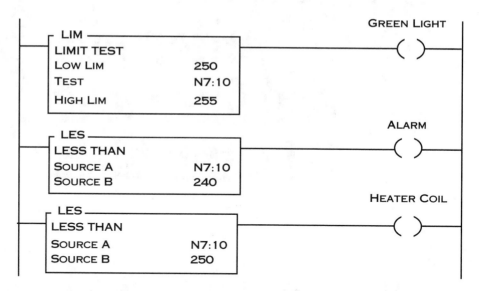

Figure 8–1 Comparison instructions used to program a simple application. In this application an integer (N7:10) contains the current temperature. The instructions are being used to compare the current temperature (N7:10) to process limits. In the first rung, the green indicator lamp is on if the temperature is between 250 and 255 degrees. In the second rung, an alarm sounds if the temperature drops below 240 degrees. In the third rung, the heater coil will be on if the temperature is below 250 degrees.

In many cases numbers need to be manipulated, to be added, subtracted, multiplied, or divided. PLC instructions can handle all of these manipulations. For example, we might want to change a system for furnace temperature control from a system to control the Celsius temperature to degrees Fahrenheit for display to the operator. If the temperature rises too high, an alarm is triggered. Arithmetic instructions could do this very easily. An example of this is shown later in this chapter.

ROCKWELL AUTOMATION ARITHMETIC INSTRUCTIONS

A multitude of arithmetic instructions are available in the PLC-5 and SLC-500, and Micrologix 1000 controllers. This section discusses several of them. Figure 8-2 shows the arithmetic functions that are available.

Add

The add instruction (ADD) adds two values together (Source A + Source B) and puts the result in the destination address. Figure 8-3 is an example of an ADD in-

Instruction	Function	Description of Operation and Use
ADD	Add	Adds source A to source B and stores the result in the destination.
SUB	Subtract	Subtracts source B from source A and stores the result in the destination.
MUL	Multiply	Multiplies source A by source B and stores the result in the destination.
DIV	Divide	Divides source A by source B and stores the result in the destination and the math register.
DDV	Double Divide	Divides the contents of the math register by the source and stores the result in the destination and the math register.
CLR	Clear	Sets all bits of a word to zero.
SQR	Square Root	Calculates the square root of the source and places the integer result in the destination.
SCP	Scale with parameters	Produces a scaled output value that has a linear relationship between the input and scaled values.
SCL	Scale Data	Multiplies the source by a specified rate, adds to an offset value, and stores the result in the destination.
ABS	Absolute	Calculates the absolute value of the source and places the result in the destination.
CPT	Compute	Evaluates an expression and stores the result in the destination.
SWP	Swap	Swaps the low and high bytes of a specified number of words in a bit, integer, ASCII, or string file.
ASN	Arc Sine	Takes the arc sine of a number (source in radians) and stores the result (in radians) in the destination.
ACS	Arc Cosine	Takes the arc cosine of a number (source in radians) and stores the result (in radians) in the destination.
ATN	Arc Tangent	Takes the arc tangent of a number (source) and stores the result (in radians) in the destination.
COS	Cosine	Takes the cosine of a number (source in radians) and stores the result in the destination.
LN	Natural Log	Takes the natural log of the value in the source and stores the result in the destination.
LOG	Log to the Base 10	Takes the log base 10 of the value in the source and stores the result in the destination.
SIN	Sine	Takes sine of a number (source in radians) and stores the result in the destination.
TAN	Tangent	Takes the tangent of a number (source in radians) and stores the result in the destination.
XPY	X to the power of Y	Raises a value to a power and stores the result in the destination.

Figure 8–2 Some Rockwell Automation math instructions.

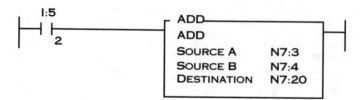

Figure 8–3 ADD instruction.

struction. If contact I:5/2 is true, the ADD instruction adds the number from source A (N7:3) and the value from source B (N7:4) and stores the result (N7:20) in the destination address.

Subtraction

The subtraction instruction (SUB) is used to subtract two values, such as source B from source A. The result is stored in the destination address. Figure 8-4 shows subtract instruction (SUB). If contact I:5/2 is true, the SUB instruction is executed. Source B is subtracted from source A; the result is stored in destination address N7:20.

Multiply

The multiply instruction (MUL) is used to multiply two values, in this case source A by source B. The result is stored in the destination address. Source A and source B can be either values or addresses of values. Figure 8-5 shows the use of a multiply instruction. If contact I:5/2 is true, source A (N7:3) is multiplied by source B (N7:4), and the result is stored in destination address N7:20.

Divide

The divide instruction (DIV) divides two values. Source A is divided by source B, and the result is placed in the destination address. The sources can be values or addresses of values. Figure 8-6 shows the use of a divide instruction. If contact

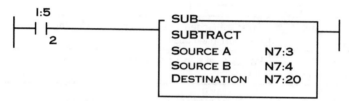

Figure 8–4 Subtract (SUB) instruction.

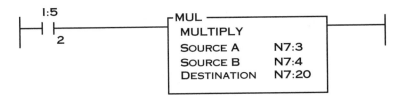

Figure 8–5 Multiply (MUL) instruction.

I:5/2 is true, the divide instruction divides the value from source A (N7:3) by the value from source B (N7:4). The result is stored in destination address N7:20.

Negate

The negate instruction (NEG) changes the sign of a value. If used on a positive number, it makes it a negative number. If it is used on a negative number, it changes it to a positive number. Remember that this instruction executes every time the rung is true. The use of a negate instruction is shown in Figure 8-7. If contact I:5/2 is true, the value in source A (N7:3) is given the opposite sign and stored in destination address N7:20.

Square Root

The square root instruction (SQR) finds the square root of a value and stores the result in a destination address. The source can be a value or the address of a value. Figure 8-8 shows the use of a square root instruction. If contact I:5/2 is true, the SQR instruction finds the square root of the value of the number found at the source address F8:3 and stores the result at destination address F8:20.

Average Instruction

The average instruction (AVE) is a file instruction used to find the average of a set of values. It calculates the average using a floating point regardless of the type specified for the file or destination. If an overflow occurs, the CPU aborts the calculation. In

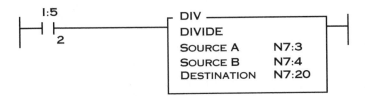

Figure 8–6 Divide (DIV) instruction.

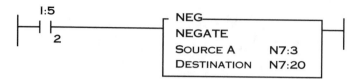

Figure 8–7 Negate (NEG) instruction.

that case the destination remains unchanged. The position points to the element that caused the overflow. When the .ER bit is cleared, the position is reset to zero, and the instruction is recalculated. Every time there is a low-to-high transition, the value of the current element is added to the next element. The next low-to-high transition causes the current element value to be added to the next element, and so on. Every time another element is added, the position field and the status word are incremented.

Figure 8-9 shows the use of an average instruction (AVE) in a ladder diagram. The programmer must provide several inputs. The file is the address of the first element to be added and used in the calculation. The destination is the address where the result will be stored. This address can be a floating point or an integer. The control is the address of the control structure in the control area (R) of CPU memory. The CPU uses this information to run the instruction. Length is the number of elements to be included in the calculation (0 to 1000). Position points to the element that the instruction is currently using.

Compute

The compute instruction (CPT) can copy arithmetic, logical, and number conversion operations. The user defines the operations to be performed in the Expression; the result is written in the Destination. Operations are performed in a prescribed order; those of equal order are performed left to right. Figure 8-10 shows the order in which operations are performed. The programmer can override precedence order by using parentheses.

Figure 8-11 is an example of a CPT instruction. The mathematical operations are performed when contact I:5/2 is true. In this case, two floating-point numbers (F8:1 and F8:2) are added to each other. Then the square root is calculated and stored in the destination address (F8:10).

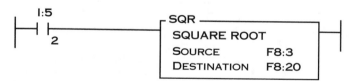

Figure 8–8 Square root (SQR) instruction.

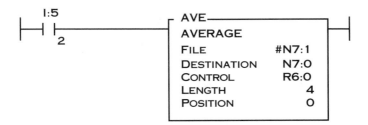

Figure 8–9 Average (AVE) instruction.

Order	Operation	Description
1	**	Exponential (x to the power of y)
2	-	Negate
	NOT	Bitwise Complement
3	*	Multiply
	/	Divide
4	+	Add
	-	Subtract
5	AND	Bitwise AND
6	XOR	Bitwise Exclusive OR
7	OR	Bitwise OR

Figure 8–10 Precedence (order) in which math operations are performed. When precedence is equal, the operations are performed left to right. Parentheses can be used to override the order.

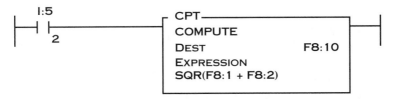

Figure 8–11 Compute (CPT) instruction.

Instruction	Function	Description of Operation and Use
EQU	Equal	The equal instruction is used to test whether two values are equal.
NEQ	Not Equal	The equal instruction is used to test whether two values are not equal.
LES	Less Than	The LES instruction is used to test whether one value (source A) is less than another (source B).
LEQ	Less Than or Equal	The LEQ instruction is used to test whether one value (source A) is less than or equal to another (source B).
GRT	Greater Than	The GRT instruction is used to test whether one value (source A) is greater than another (source B).
GEQ	Greater Than or Equal	The GEQ instruction is used to test whether one value (source A) is less than or equal to another (source B).
MEQ	Masked Comparison for Equal	The MEQ instruction is used to compare data at a source address with data at a compare address.
LIM	Limit Test	The limit instruction is used to test for values within or outside a specified range.

Figure 8–12   Rockwell Automation relational operators.

Relational Operators

Rockwell Automation also has several comparison-type instructions available. Rockwell calls them relational operators. The instructions can be used to compare two numbers. See Figure 8-12 for a list of Rockwell Automation relational operator instructions.

Equal To   The equal to instruction (EQU) tests whether two values are equal for actual values or addresses that contain values. See Figure 8-13 for an example. Source A is compared to source B to test to determine whether they are equal. If the value in N:7 equals the value in N7:10, the rung is true, and output O:5/01 is turned on.

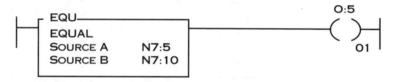

Figure 8–13   Equal to instruction (EQU).

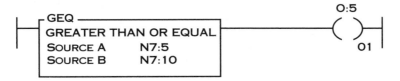

Figure 8–14   A greater than or equal to (GEQ) instruction.

Greater Than or Equal To   The greater than or equal to instruction (GEQ) tests two sources to determine whether source A is greater than or equal to the second source. A GEQ instruction is shown in Figure 8-14. If the value of source A (N7:5) is greater than or equal to source B (N7:10), output O:5/01 is turned on.

Greater Than   The greater than instruction (GRT) determines whether a value from one source is greater than the value from a second source. An example of the instruction is shown in Figure 8-15. If the value of source A (N7:5) is greater than the value of source B (N7:10), output O:5/01 is set (turned on).

Less Than   The less than instruction (LES) determines whether a value from one source is less than the value from a second source. See Figure 8-16 for an example of the instruction. If the value of source A (N7:5) is less than the value of source B (N7:10), output O:5/01 is set (turned on).

Limit   The limit instruction (LIM) tests a value to see whether it falls in a specified range of values. The instruction is true when the tested value is within the limits. This could be used, for example, to determine whether the temperature of an oven is within the desired temperature range. In this case, the instruction tests an analog value (a number in memory representing the actual analog temperature).

The programmer must provide three pieces of data to the LIM instruction when programming: a low limit, a test value, and the high limit. The low limit can be a constant or an address that contains the desired value; if it is an address it contains

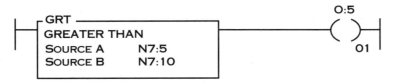

Figure 8–15   Greater than (GRT) instruction.

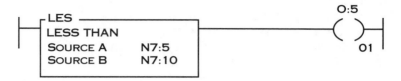

Figure 8–16 Less than (LES) instruction.

an integer or floating-point value (16 bits). The test value is a constant or the address of a value to be tested. If the test value is within the range specified, the rung is true. The high limit can be a constant or the address of a value.

Figure 8-17 shows the use of a limit (LIM) instruction. If the value in N7:15 is greater than or equal to the lower limit value (N7:10) and less than or equal to the high limit (N7:20), the rung is true and output O:5/01 is turned on.

Not Equal To The not equal to instruction (NEQ) tests two values (constants or addresses) that contain the values for inequality. In Figure 8-18, if source A (N7:3) is not equal to source B (N7:4), the instruction is true and output O:5/01 is turned on.

LOGICAL OPERATORS

Several logical instructions that can be very useful to the innovative programmer are available. They can be used, for example, to check the status of certain inputs while ignoring others.

And

One logical operator instruction is the AND instruction used to perform an AND operation using the bits from two-source addresses. The bits are "ANDed" and a result occurs. See Figure 8-19 for a chart that shows the results of the four possible combinations. These two sources (numbers) are ANDed, and the result is stored in a third address (see Figure 8-20). See Figure 8-21 for an example of an

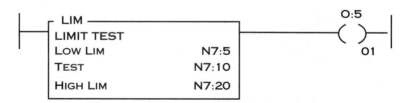

Figure 8–17 Limit test (LIM) instruction.

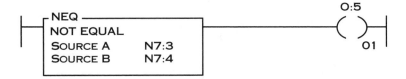

Figure 8–18 Use of a not equal to (NEQ) instruction.

Figure 8–19 Results of an AND operation on the four possible bit combinations.

Source A	Source B	Result
0	0	0
1	0	0
0	1	0
1	1	1

Source A N7:3	0	0	0	0	0	0	0	0	1	0	1	0	1	0	1	0
Source B N7:4	0	0	0	0	0	0	0	0	1	1	1	0	1	0	1	1
Destination N7:5	0	0	0	0	0	0	0	0	1	0	1	0	1	0	1	0

Figure 8–20 Result of an AND on two source addresses. The ANDed result is stored in address N7:5.

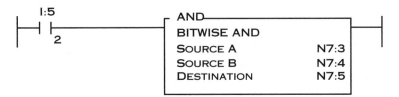

Figure 8–21 AND instruction: If input I:5/2 is true, the AND instruction executes. The number in address N7:3 is ANDed with the value in address N7:4. The result of the AND is stored in address N7:5.

Figure 8–22 Results of a NOT instruction on bit states.

Source	Result
0	1
1	0

AND instruction. Addresses N7:3 and address N7:4 are ANDed. The result is placed in destination address N7:5. Examine the bits in the source addresses to understand how the AND produced the result in the destination.

NOT

NOT instructions are used to invert the status of bits such as making 1 a 0 and a 0 a 1. See Figures 8-22, 8-23, and 8-24.

OR

Bitwise OR instructions are used to compare the bits of two numbers. See Figures 8-25, 8-26, and 8-27 for examples of an OR instruction's result, its use in a ladder diagram, and another result.

GENERAL EXPLANATION OF SEQUENTIAL CONTROL

Before PLCs, many innovative methods were used to control machines. One of the earliest control methods was the punched card. These date back to the use in the earliest automated weaving machines to control the weave. Until a relatively few years ago, the main method of input to computers was through punched cards.

Most manufacturing processes are very *sequential*, meaning that they process a series of steps, from one to the next, such as a bottling line. Bottles enter the line, are cleaned, filled, capped, inspected, and packed. In addition, many home appli-

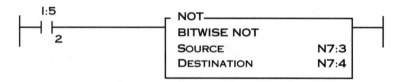

Figure 8–23 NOT instruction: If input I:5/02 is true, the NOT instruction executes. The number in address N7:3 is NOTed (1 is complemented). The result is stored in destination address N7:4.

Source A N7:3	0	0	0	0	0	0	0	0	1	0	1	0	1	0	1	0
Destination N7:4	0	0	0	0	0	0	0	0	0	1	0	1	0	1	0	1

Figure 8–24 What would happen to the number 0000000001010101010 if a NOT instruction were executed? The result is shown in destination address N7:4.

Figure 8–25 Result of an OR instruction on bit states.

Source A	Source B	Result
0	0	0
1	0	1
0	1	1
1	1	1

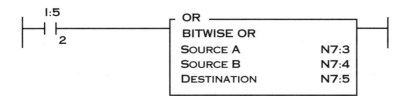

```
I:5                    ┌─ OR ─────────────────────┐
─┤ ├──────────────     │ BITWISE OR               │──────
   2                   │ SOURCE A          N7:3   │
                       │ SOURCE B          N7:4   │
                       │ DESTINATION       N7:5   │
                       └──────────────────────────┘
```

Figure 8–26 OR instruction used in a ladder diagram. If input I:5/2 is true, the OR instruction executes. Source A (N7:3) is "ORed" with source B (N7:4). The result is stored in the destination address.

Source A N7:3	0	0	0	0	0	0	0	0	1	0	1	0	1	0	1	0
Source B N7:4	0	0	0	0	0	0	0	0	1	1	1	0	1	0	1	1
Destination N7:5	0	0	0	0	0	0	0	0	1	1	1	0	1	0	1	1

Figure 8–27 Result of an OR instruction on the numbers in address N7:3 and address N7:4. The result is shown in the destination address N7:5.

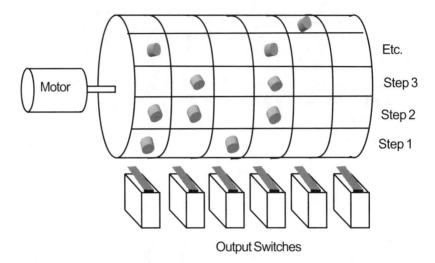

Etc.

Step 3

Step 2

Step 1

Output Switches

Figure 8–28   Drum controller. Note the pegs that activate switches as the motor turns the drum at slow speed. See Figure 8-29, which shows the output conditions for the steps.

ances work sequentially, including washers, dryers, dishwashers, and breadmakers. Plastic injection molding, metal molding, packaging, and filling are several sequential industrial processes.

Many of these machines were (and some still are) controlled by a device called a *drum controller*. It functions just like an old player piano, which was controlled by a paper roll with holes punched in it. The position of the holes across the roll indicates which note should be played. Their position around the roll indicates when they are to be played. A drum controller is the player piano's industrial equivalent.

The drum controller is a cylinder with holes around the perimeter with pegs placed in the holes (see Figure 8-28). The pegs hit switches as the drum turns. The peg turns, closing the switch that it contacts and turning on the output to which it is connected. A motor controls the speed of the drum and each step must take the same amount of time. If an output must be on longer than one step, consecutive pegs must be installed.

The drum controller has several advantages. It is easy to understand, which makes it easy for a plant electrician to work with. It is easy to maintain. It is easy to program; the user makes a simple chart that shows which outputs are on in which steps (see Figure 8-29). The user then installs the pegs to match the chart.

Step	Input Pump	Heater	Add Cleaner	Sprayer	Output Pump	Blower
1	on		on			
2	on	on		on		
3		on		on		
4					on	
5						on

Figure 8–29 Output conditions for the drum controller shown in Figure 8-28.

Many home appliances are also controlled with drum technology. Instead of a cylinder, they utilize a disk with traces (see Figure 8-30). Think of a washing machine. The user chooses the wash cycle by setting the dial to the proper position. The disk is then moved very slowly as a synchronous motor turns. Contacts (brushes) make contact with traces at the proper times and turn on output devices such as pumps and motors.

Although this type of control has many advantages, it also has some major limitations. The time for individual steps cannot be controlled individually. The sequence is set and continues even if something goes wrong. The drum continues to turn, turning devices on and off. In other words, the step occurs even if certain conditions are not met. PLC instructions have been designed to include the drum controller's strengths and to overcome its weaknesses.

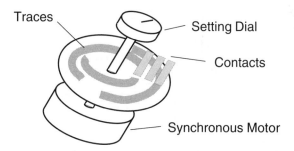

Figure 8–30 Disk-type drum controller commonly used in home appliances. Turning the setting dial actually turns the disk to its starting position.

Sequencer Instructions

Sequencer instructions can be used to monitor inputs to control the sequencing of the outputs of any process that is cyclical in nature. Like the drum controller, a sequencer instruction can make programming many applications a much easier task.

Specific Explanation of Sequencer Instructions

Sequencer instructions can easily program processes with defined steps. Rockwell Automation PLCs have three main instructions: sequencer output (SQO), sequencer compare (SQC), and sequencer load (SQL). Imagine a sequential process that has six steps. We could write down the states of the outputs for every step (see Figure 8-31). Note that in step 1, output 0 is on. In step 2, output 0 stays on and output 1 turns on. This would be a good application for a sequencer output instruction (SQO). Figure 8-32 shows a simple ladder diagram with a timer and an SQO instruction.

The first entry in the SQO is the file number to be used. File is the starting location in memory for output conditions for each step. In this example, the starting file address is N7:0. The output states for each step are entered next starting at N7:0 (see Figure 8-33). This process has seven steps, located in words N7:0, N7:1, N7:2, N7:3, N7:4, N7:5, and N7:6. The first word is used for step 0, which is not actually part of the six operational steps. SQOs always start at step 0 the first time. When the SQO reaches the last step, however, it resets to step 1 (position 1).

Next we would enter a mask, which can be a file in memory or one word (16 bits) in memory. If an address is entered for the mask value, the mask will be a list of values starting at that address. For example, if the user enters N7:20, N7:20 through N7:25 are the locations of the mask values for each of the six steps. The user would have to fill these locations with the desired mask values. In this example, each of the six mask values would correspond to one step.

A constant can also be used. If a constant is entered, the SQO uses the constant as a mask for all of the steps. In our example a hex number was entered (see

Step #	Output 5	Output 4	Output 3	Output 2	Output 1	Output 0
1						On
2					On	On
3				On		On
4			On			
5		On		On		
6	On			On		

Figure 8–31 Outputs that are on in each step.

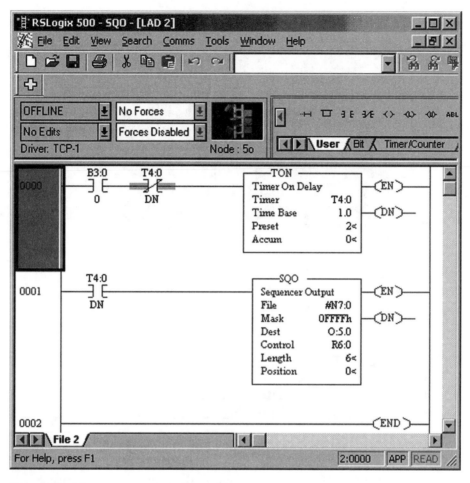

Figure 8–32 Ladder logic with a timer and an SQO instruction.

Figure 8-32). The four hex digits represent 16 bits. Each bit corresponds to one output in our SQO. If the bit is a 1 in the mask position, the output is enabled and turns on if the step condition tells it to be on. In other words, the bit condition for the active step is ANDed with the mask. If the bit in the step is a 1 and the corresponding mask bit is a 1, the output is set.

The destination is the location of the outputs. In this example the outputs are in O:5.0 (the output module in slot 5) meaning that output 0 is O:5/0, output 1 is O:5/1, and so on. Remember that the output states for each step are located in N7:1 through N7:6. When the ladder diagram is in the first step, output 0 is on. In the next step, outputs 0 and 2 are on.

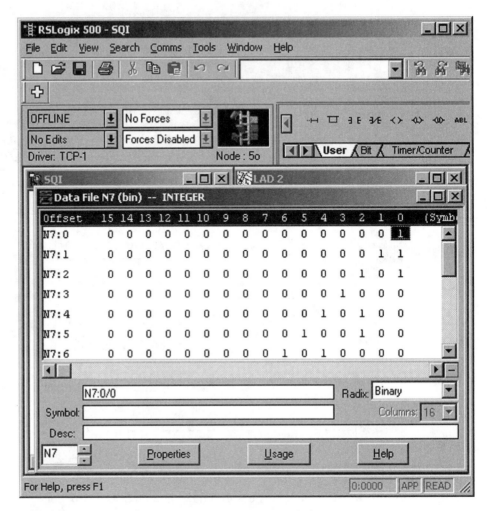

Figure 8–33 Desired output conditions stored in memory.

A file location must be entered for control (see Figure 8-32). Control is where the SQO stores status information for the instruction. An SQO or SQC uses three words in memory (see Figure 8-34). Three bits can be used: EN—enable, DN—done, and the ER—error bit. Word 1 contains the length of the sequencer (number of steps). Word 2 contains the current position (step in the sequence).

Length is the number of steps in the process starting at position 1. The maximum length is 255 words. Position 0 is the startup position. The first time the SQO is enabled, it moves from position 0 to position 1. The instruction resets to position 1 at the end of the last step. Position is the word location (or step) in the sequencer file from/to which the instruction moves data. In other words, position shows the

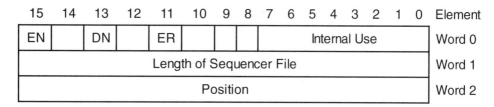

Figure 8–34 Memory organization for an SQO control memory.

number of the step that is currently active. Let's see how it would work. First, assume that this process is very simple and every step takes 5 seconds.

A 5-second timer is used as the input condition to the SQO instruction. Every time the timer's accumulated value reaches 5 seconds, the timer done bit forces the SQO instruction to move to the next step. That step's outputs are turned on for 5 seconds until the timer done bit enables the SQO again. The SQO then moves to the next step and sets any required outputs. Note that the timer done bit also resets the accumulated time in the timer to zero and the timer starts timing to 5 seconds again. When the SQO is in the last step and receives the enable again, the SQO returns to step 1. (The first time the ladder is executed, the SQO starts at step 0).

In the example SQO we just examined, every step was the same length of time. In most applications, however, specific input conditions need to be met before we can move to the next step of the process. We can use any logic we would like to enable the SQO. It can get quite complex to set up all of the different conditions in logic. Fortunately, Rockwell Automation PLCs have a sequential compare (SQC) instruction that can make our task easy. If we make a list of the input conditions we need for every step, we can use those as a way to trigger the SQO (see Figure 8-35).

Step	Input 3	Input 2	Input 1	Input 0
1		On		On
2		On	On	
3	On			
4			On	
5		On		
6	On			On

Figure 8–35 Input condition table. Note that if we are presently in step 2, we need inputs 1 and 2 to be true in order to move to step 3.

Note that the SQC instruction in Figure 8-32 had to give the starting file number as the starting address. In this example, N7:10 was entered as the starting address. The desired input conditions to move from step 1 to step 2 are entered into N7:11 (see Figure 8-36). Remember that the sequencer initially begins with step 0, the actual program steps begin with step 1. The remaining desired input and conditions are entered as shown in Figure 8-36.

Figure 8-37 shows a SQC. File is the location of the first word of input conditions. In our example, it is a six-step process, so we would have six input words. The mask is used as it is with the SQO. The mask can be a file in memory or one word (16 bits) in memory. If an address is entered for the mask value, the mask is a list of values starting at the address specified. For example, if the user enters

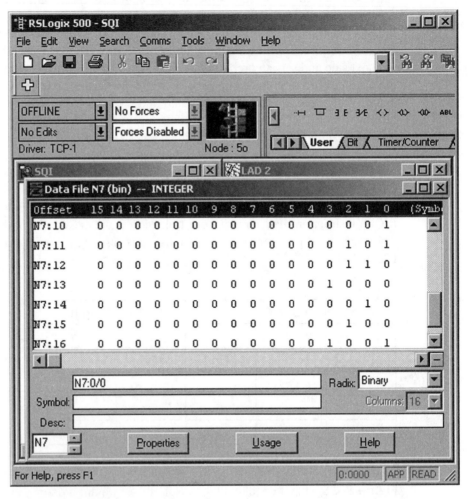

Figure 8-36 Desired input conditions in integer memory.

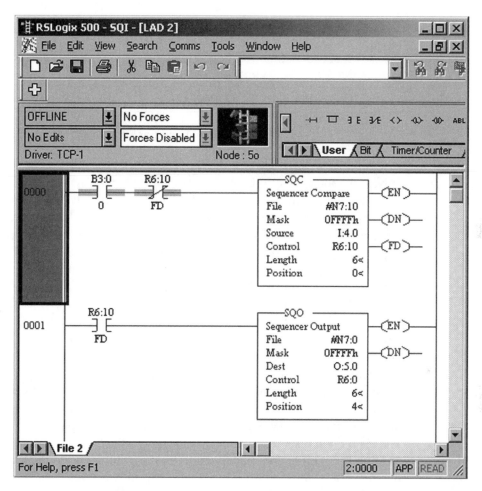

Figure 8–37   Ladder logic showing the use of a SQC and an SQO.

N7:30, N7:30 through N7:35 are the location of the mask values for each of the six steps. The user must fill these locations with the desired mask values. Each mask value corresponds to a step in the sequence. This example has six steps, so each of the six mask values corresponds to one step.

Otherwise a constant can be used. In our case a hex constant was entered. The four hex digits correspond to 16 bits. If the bit is a 1 in the mask, that input is enabled. The source word (usually real-world input conditions) is ANDed with the mask value and compared to the current step (word) in the file. In this case our input mask is 000FH. In this application we are interested in the states of only the first 4 inputs. The 000FH (H = HEX) means the first four inputs are used and that the last 16 are ignored (0000 0000 0000 1111).

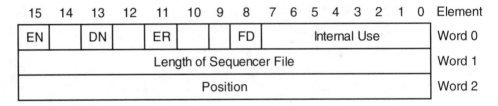

Figure 8–38   Control register memory address for a SQC instruction.

Next the source, the address of the real-world inputs, must be entered. In this example the inputs are at I:4.0. The address I:4.0 is 16 inputs in the module in slot 4. The states of these real-world inputs are compared to the desired inputs in the present step. If they are the same, the SQC position value is incremented to the next step and the FD bit is set in the control register. The FD bit (R6:21/FD in this case) triggers the SQO to transition to the next step. This means that the desired input conditions can be used to control the transition from step to step in our process.

A file location must also be entered for control (Figure 8-38). Control is the location where the SQC stores status information for the instruction. An SQC uses three words in memory (see Figure 8-38). Four bits can be used: EN—enable, DN—done, ER—error, and FD—found bit. Word 1 contains the length of the sequencer (number of steps). Word 2 contains the current position.

Length, which is the number of steps in the process starting at position 1, must also be entered. The maximum length is 255 words. Position 0 is the startup position. The first time the SQO is enabled, it moves from position 0 to position 1. The instruction resets to position 1 at the end of the last step. *Position* is the word location (or step) in the sequencer file from/to which the instruction moves data. In other words, position is the step that is currently active.

Sequencer Load Instruction

The sequencer load instruction (SQL) can be used to store source information (words) into memory. For example, input conditions could be stored into integer memory every time the sequencer load instruction sees a false-to-true transition. When it gets to the last position, it will transition back to step 1 when it sees the last false-to-true transition.

SHIFT RESISTER PROGRAMMING

A *shift register* is a storage location in memory that can typically hold 16 bits of data, that is, 1s or 0s. Each 1 or 0 could be used to represent good or bad parts, presence or absence of parts, or the status of outputs (see Figure 8-39). Many

Station #	1	2	3	4	5	6	7	8
Part Present	1	0	0	0	1	0	0	1

Figure 8–39 A shift register might look like this when monitoring whether or not parts are present at processing stations. A 1 in the station location is used to turn an output on and cause the station to process material. In this case, the PLC turns on outputs at stations 1, 5, and 8. After processing, all bits shift to the right. A new 1 or a 0 is loaded into the first bit, depending on whether a part was or was not present. The PLC then turns on all stations that had a 1 in their bit. In this way, processing occurs only when parts are present.

manufacturing processes are very linear in nature; a bottling line referred to previously is an example. The bottles are cleaned, filled, capped, and so on. The sensors along the way sense for the presence of a bottle, check the fill, and so on. These conditions could easily be represented by 1s and 0s.

Shift registers essentially shift bits through a register to control I/O. The bottling line in the example has many processing stations, each represented by a bit in the shift register. We want to run the processing station only if parts are present. As the bottles enter the line, a 1 is entered into the first bit. Processing takes place. Each station then releases its product which moves to the next station. The shift register also increments. Each bit is shifted one position. Processing takes place again. Each time a product enters the system, a 1 is placed in the first bit. The 1 follows the part all the way through production to make sure that each station processes it as it moves through the line. Shift register programming is very applicable to linear processes.

QUESTIONS

1. Explain some of the reasons why arithmetic instructions are used in ladder logic.

2. For what are LIM instructions used?

3. For what are comparison instructions used?

4. Why might a programmer use an instruction that would change a number to a different number system?

5. Write a rung of ladder logic to compare two values to see whether the first is greater than the second. Turn an output on if the statement is true.

6. Write a rung of logic that checks to see whether one value equals a second value. Turn on an output if true.

Figure 8–40 Tank level
application.

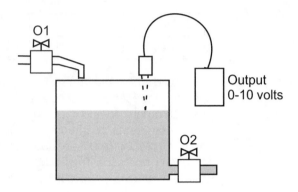

7. Write a rung of logic that checks to see whether a value is less than 20 or greater than 40. Turn on the output if the statement is true.

8. Write a rung of logic to check whether a value is less than or equal to 99. Turn on an output if the statement is true.

9. Write a rung of logic to check whether a value is less than 75, greater than 100, or equal to 85. Turn on an output if the statement is true.

10. Write a ladder logic program that accomplishes the following. A production line produces items packaged 12 to a pack. Your boss asks you to modify the ladder diagram so that the number of items is counted and the number of packs is counted. A sensor senses each item as it is produced. Use the sensor as an input to the instructions you will use to complete the task. (*Hint:* One way is to use a counter and at least one arithmetic statement.)

11. Write a ladder diagram program to accomplish the following. A tank level must be maintained between two levels (see Figure 8-40). An ultrasonic sensor is used to measure the height of the fluid in the tank. The output from the ultrasonic sensor is 0 to 10 volts (see Figure 8-41). This directly relates to a tank level of 0 to 5 feet. It is desired that the level be maintained between 4.0 and 4.2 feet. Output 1 is the input valve. Output 2 is the output valve. The sensor output is an analog input to an analog input module.

PROGRAMMING EXERCISES

These exercises utilize the LogixPro software included with this book. Utilize the LogixPro Student Exercises and Documentation. Remember the Rockwell Automation Instruction Set Reference, which may be very helpful.

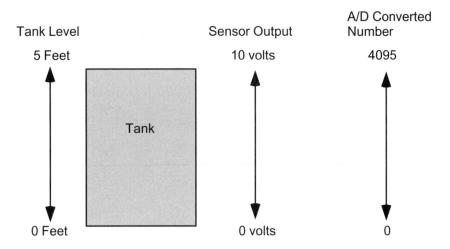

Figure 8–41 A comparison of tank level and A/D output.

1. Complete the Word Compare Introductory exercise in LogixPro.
2. Complete Exercise 1—Traffic Control utilizing one timer in LogixPro software.

EXTRA CREDIT

Complete Exercise 2—Dealing with Pedestrians in LogixPro software.

chapter

Industrial Sensors

This chapter examines the types and uses of industrial sensors. It discusses digital and analog sensors and wiring of sensors.

OBJECTIVES

Upon completion of this chapter, you will be able to:
1. Describe at least two ways in which sensors can be classified.
2. Choose an appropriate sensor for a given application.
3. Describe the typical uses of digital sensors.
4. Describe the typical uses of analog sensors.
5. Explain common sensor terminology.
6. Explain the wiring of load and line-powered sensors.
7. Explain how field sensors function.
8. Explain the principle of operation of thermocouples.
9. Explain common thermocouple terms.

THE NEED FOR SENSORS

Sensors have become vital in industry, and manufacturers are moving to integrating pieces of computer-controlled equipment. In the past, operators were the brains of the equipment as the source of all information about the operation of a process. The operator knew whether parts were available, which parts were ready, whether they were good or bad, the tooling was acceptable, the fixture was open or closed, and so on. The operator could sense problems in the operation by seeing, hearing, feeling (vibration, etc.), and smelling problems.

Industry is now using computers (in many cases PLCs) to control the motions and sequences of machines. PLCs are much faster and more accurate than an operator at these tasks. PLCs cannot see, hear, feel, smell, or taste processes by themselves but use industrial sensors to give industrial controllers these capabilities.

The PLC can use simple sensors to check whether parts are present or absent, to size the parts, and even to check if the product is empty or full. The use of sensors to track processes is vital for the success of the manufacturing process and to ensure the safety of the equipment and operator. In fact, sensors perform simple tasks more efficiently and accurately than people do. Sensors are much faster and make far fewer mistakes.

Studies have been performed to evaluate how effective human beings are in doing tedious, repetitive inspection tasks. One study examined people inspecting table tennis balls. A conveyor line brought table tennis balls by a person. White balls were considered good, and black balls were considered scrap. The study found that people were only about 70 percent effective at finding the defective ping pong balls. Certainly, people can find all of the black balls, but they do not perform mundane, tedious, repetitive tasks well; they become bored and make mistakes, whereas a simple sensor can perform simple tasks almost flawlessly.

SENSOR TYPES

Contact vs. Noncontact

Sensors are classified in a number of ways; one common classification is contact or noncontact. The use of a noncontact sensor, also called a *proximity sensor,* is a simple way to identify a sensor. If the device must contact a part to sense it, the device is a *contact sensor.* A simple limit switch on a conveyor is an example. When the part moves the lever on the switch, the switch changes state. The contact of the part and the switch creates a change in state that the PLC can monitor.

Noncontact sensors can detect the part without touching it physically, which avoids slowing down or interfering with the process. *Noncontact sensors* (electronic) do not operate mechanically (i.e., they have no moving parts) and are more reliable and less likely to fail than mechanical ones. Electronic devices are also much faster than mechanical devices, so noncontact devices can perform at very high production rates.

The remainder of the chapter examines noncontact sensors.

Digital Sensors

Another way to classify sensors is as digital or analog. Industrial applications need both digital and analog sensors. A *digital sensor* has two states: on or off. Most applications involve presence/absence and counting, which a digital sensor does perfectly and inexpensively. Digital sensors are simpler and easier to use than analog ones, which is a factor in their wide use. Computers are digital devices that actually work with only 1s or 0s (on or off).

Digital output sensors are either on or off. They generally have transistor outputs. If the sensor senses an object, the transistor turns on and allows current to flow. The output from the sensor is usually connected to a PLC input module.

As discussed in Chapter 4, sensors are available with either normally closed or normally open output contacts. Normally open contact sensors are off until they sense an object and turn on. Normally closed contact sensors are on until they sense an object, when they turn off. When photosensors are involved, the terms light-on and dark-on are often used. *Dark-on* means that the sensor output is on when no light returns to the sensor, which is similar to a normally closed condition. A *light-on* sensor's output is on when light returns to the receiver, similar to a normally open sensor.

The current limit for most sensors' output is quite low. Usually output current must be limited to less than 100 milliamps. Users should check the sensor before turning on the power. Output current must be limited or the sensor can be destroyed! This is usually not a problem if the sensor is being connected to a PLC input since the PLC input limits the current to a safe amount.

Analog sensors, also called *linear output sensors*, are more complex than digital ones but can provide much more information about a process.

Think about a sensor used to measure the temperature. A temperature is analog information. The temperature in the Midwest is usually between 0 and 90 degrees. An analog sensor could sense the temperature and send a current to the PLC. The higher the temperature, the higher the output from the sensor. The sensor may, for example, output between 4 and 20 milliamps, depending on the actual temperature, although there are an unlimited number of temperatures (and thus current outputs). Remember that the output from the digital sensor is on or off. The output from the analog sensor can be any value in the range from low to high. Thus, the PLC can monitor temperature very accurately and control a process closely. Pressure sensors are also available in analog style. They provide a range of output voltage (or current), depending on the pressure.

A 4-20 milliamp current loop system can be used for applications when the sensor needs to be mounted a long distance from the control device. A 4-20 milliamp loop is good for about 800 meters. A 4-20 milliamp sensor varies its output between 4 and 20 milliamps. The sensor must be adjusted for range and sensitivity so that it can measure the required value of the characteristic, such as temperature.

DIGITAL SENSORS

Digital sensors come in many types and styles, which are examined next.

Optical Sensors

All optical sensors use light to sense objects in approximately the same manner. A light source (emitter) and a photodetector sense the presence or absence of light. Light-emitting diodes (LEDs), which are semiconductor diodes that emit light, are typically used for the light source because they are small, sturdy, very efficient, and can be turned on and off at extremely high speeds. They operate in a narrow wavelength and are very reliable. LEDs are not sensitive to temperature, shock, or vibration and have an almost endless life. The type of material used for the semiconductor determines the wavelength of the emitted light.

The LEDs in sensors are used in a pulsed mode. The photo emitter is pulsed (turned off and on repeatedly). The "on time" is extremely small compared to the "off time." LEDs are pulsed for two reasons: to prevent the sensor from being affected by ambient light, and to increase the life of the LED. This can also be called *modulation*.

The photodetector senses the pulsed light. The photo emitter and photo receiver are both "tuned" to the frequency of the modulation. The photodetector essentially sorts out all ambient light and looks for only the correct frequency. Light sources chosen are typically invisible to the human eye and wavelengths are chosen so that the sensors are not affected by other lighting in the plant. The use of different wavelengths allows some sensors, called *color mark sensors*, to differentiate between colors. Visible sensors are usually used for this purpose. The pulse method and the wavelength chosen make optical sensors very reliable.

Some sensor applications utilize ambient light emitted by red-hot materials such as glass or metal. These applications can use photo receivers sensitive to infrared light.

All of the various types of optical sensors function in the same basic manner. The differences are in the way in which the light source (emitter) and receiver are packaged.

Light/Dark Sensing   Optical sensors are available in either light or dark sensing, also called *light-on* or *dark-on*. In fact, many sensors can be switched between light and dark modes. Light/dark sensing refers to the normal state of the sensor and whether its output is on or off in its normal state.

Light Sensing (Light-on)   The output is energized (on) when the sensor receives the modulated beam of light. In other words, the sensor is on when the beam is unobstructed.

Dark Sensing (Dark-on) The output is energized (on) when the sensor does not receive the modulated beam. In other words, the sensor is on when the beam is blocked. Street lights are examples of dark-on sensors. When it gets dark outside, the street light is turned on.

Timing Functions Timing functions are available on some optical sensors. They are available with on-delay and off-delay. *On-delay* delays turning on the output by a user-selectable amount. *Off-delay* holds the output on for a user-specified time after the object has moved away from the sensor.

Types of Optical Sensors

Reflective Sensors One of the more common types of optical sensors is the *reflective* or *diffuse reflective type*. The light emitter and receiver are packaged in the same unit. The emitter sends out light, which bounces off the product to be sensed. The reflected light returns to the receiver where it is sensed (Figure 9-1). Reflective sensors have less sensing distance (range) than other types of optical sensors because they rely on light reflected off the product.

Polarizing Photo Sensor A special photo sensor senses shiny objects using a special reflector. The reflector actually consists of small prisms that polarize the light from the sensor. The reflector vertically polarizes the light and reflects it back to the sensor receiver. The photo sensor emits horizontally polarized light.

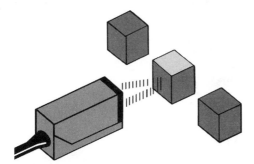

Figure 9–1 Reflective-type sensor. The light emitter and receiver are in the same package. When the light from the emitter bounces off an object, the receiver senses it and the sensor's output changes state. The broken-line style of the arrows represents the pulsed mode of lighting, which is used to ensure that ambient lighting does not interfere with the application. The sensing distance (range) of this style is limited by how well the object can reflect the light back to the receiver. (*Courtesy ifm efector inc.*)

Figure 9-2 A polarizing
photo sensor. Note the use of
the special reflector.

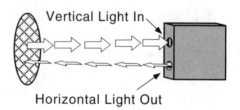

Vertical Light In

Horizontal Light Out

Thus, if a very shiny object moves between the photo sensor and reflector and reflects light back to the sensor, it is ignored because it is not vertically polarized (see Figure 9-2).

Retroreflective Sensor   The retroreflective sensor (see Figure 9-3) is similar to the reflective sensor. The emitter and receiver are both mounted in the same package. The difference is that the retroreflective sensor bounces the light off a reflector instead of the product. The reflector is similar to those used on bicycles. Retroreflective sensors have more sensing distance (range) than do reflective (diffuse) sensors but less sensing distance than that of a thru-beam sensor. They are a good choice when scanning can be done from only one side of the application, which usually occurs when space is a limitation.

Thru-Beam Sensor   Another common sensor is the thru-beam (Figure 9-4), also called the *beam break mode*. In this configuration the emitter and receiver are packaged separately. The emitter sends out light across a space and the receiver senses the light. If the product passes between the emitter and receiver, it stops

Figure 9-3 Retroreflective sensor. The light emitter and receiver are in the same package. The light bounces off a reflector and is sensed by the receiver. If an object obstructs the beam, the output of the sensor changes state. The excellent reflective characteristics of a reflector give this sensor more sensing range than a typical diffuse style does. The broken-line arrows represent the pulsed method of lighting that is used. (*Courtesy ifm efector inc.*)

Figure 9–4 Thru-beam sensor. The emitter and receiver are in separate packages. The broken-line arrow symbolizes the pulsed mode of the light that is used in optical sensors. (*Courtesy ifm efector inc.*)

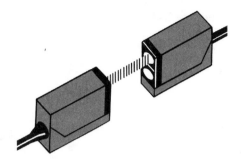

the light from hitting the receiver, telling the sensor that a product is present. This is probably the most reliable sensing mode for opaque (nontransparent) objects.

Convergent Photo Sensor   A convergent photo sensor, *also called focal length sensor*, is a special type of reflective sensor. It emits light to a specific focal point. The light must be reflected from the focal point to be sensed by the sensor's receiver (see Figure 9-5).

Fiber-Optic Sensor   A fiber-optic sensor is simply a mix of the other types. The actual emitter and receiver are the same but with a fiber-optic cable attached to each. Fiber-optic cables are very small and flexible and are transparent strands of plastic or glass fibers used as a "light pipe" to carry light. The cables are available in both thru-beam and reflective configuration (see Figure 9-6). The light enters the end of the cable attached to the receiver, passes through the cable, and is sensed by the receiver. The light from the emitter passes through the cable and exits from the other end.

Color Mark Sensor   A color mark sensor is a special type of diffuse reflective optical sensor that can differentiate between colors; some can even detect contrast between colors. It is typically used to check labels and to sort packages by color mark. A sensitivity adjust function can make fine adjustments. The object's background color is an important consideration. Sensor manufacturers provide charts for the proper selection of color mark sensors for various colors.

Figure 9–5 A convergent-style photo sensor.

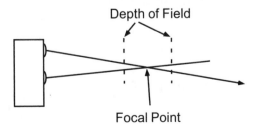

Figure 9–6 Fiber-optic sensor. (*Courtesy ifm efector inc.*)

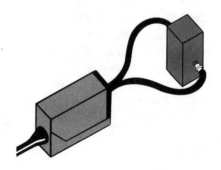

Laser Sensor A laser sensor is also used as a light source for optical sensors that performs precision-quality inspections requiring very accurate measurements. Resolution can be as small as a few microns. A laser LED is used as the source. Outputs can be analog or digital. The digital outputs can be used to signal pass/fail or other indications. The analog output can be used to monitor and record actual measurements.

Encoder Sensor An encoder sensor is used for position feedback and in some cases for velocity feedback. The two main types of encoders are incremental and absolute; incremental is the most common type. The resolution of an encoder is determined by the number of lines on the encoder disk. The more lines, the higher is the resolution. Encoder increments with 500, 1000, or even more lines are common. The light from an LED shines through the lines on the encoder disk and a mask and is then sensed by light receivers (photo transistors). See Figure 9-7 for an enlarged view of an encoder.

Incremental Encoder An incremental encoder (Figure 9-8) creates a series of square waves. Incremental encoders are available in various resolutions, which are determined by the number of slits for light to pass through. For example, a 500-count encoder produces 500 square waves in one revolution or 250 pulses with a 180-degree turn. The two main types of incremental encoders are tachometer (single-track) and quadrature (multitrack).

Tachometer Encoders Sometimes called a *single-track encoder*, a tachometer encoder has only one output and cannot detect the direction of travel. Its output is a square wave; its velocity can be determined by the frequency of pulses. The disk in Figure 9-9 shows a tachometer-style encoder disk.

Quadrature Encoder Quadrature is a configuration in which the two output channels have an angular separation of 90 degrees. Most encoders use two quadrature output channels for position sensing. Figure 9-10 shows what the A and B

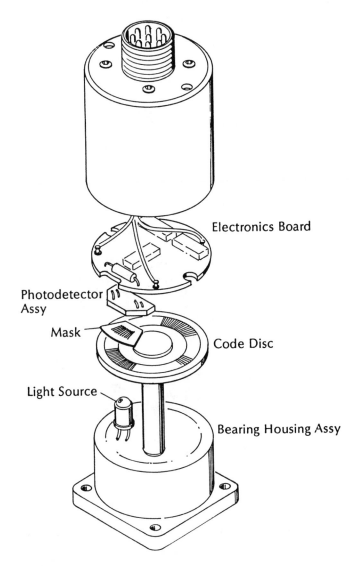

Electronics Board

Photodetector
Assy

Mask

Code Disc

Light Source

Bearing Housing Assy

Figure 9–7 An enlarged view of an encoder. (*Courtesy BEI Sensors & Systems Company.*)

pulses would look like on a oscilloscope. Note that the A and B pulses are square waves. Every slit in the encoder disk represents a high pulse. Note the relationship between the A and B pulses.

The transitions from high to low or low to high on these channels can be sensed by a computer, PLC, or special controller such as a CNC, or robot controller. Most incremental quadrature encoders have three tracks (see Figure 9-11).

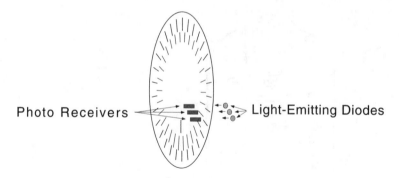

Photo Receivers   Light-Emitting Diodes

Figure 9–8   An incremental encoder.

Figure 9–9   A disk from a tachometer-style incremental encoder. (*Courtesy BEI Sensors & Systems Company.*)

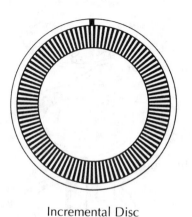

Incremental Disc

Figure 9–10   A and B ring output from the encoder on an oscilloscope.

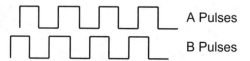

A Pulses

B Pulses

Figure 9–11   An incremental encoder.

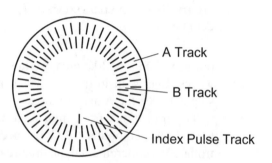

A Track

B Track

Index Pulse Track

The A the B tracks have the same number of slits. The index track has only one pulse per revolution, usually called a *zero* or *index pulse*. This is used to establish the home position for position controllers.

The A and B channels can be compared to determine the direction of rotation. By using one channel as a reference we can tell whether the transition on B leads or lags the A transition, indicating the direction of rotation. Study Figure 9-12 to see how direction can be determined by comparing the two channels.

A quadrature encoder might have a total of six lines: three for the A, B, and home pulse signals; two to supply power to the encoder; and one for case ground. A 13-bit absolute encoder has 13 output wires plus two power lines. If the absolute encoder is provided with complementary outputs, the encoder requires 28 wires.

The number of transitions identifies the amount of rotation. The quadrature encoder has a distinct advantage over the tachometer encoder for position sensing. Imagine what would happen if a tachometer encoder stopped right on a transition and the machine vibrated in such a way that the encoder continued to output transitions due to the vibration, forcing the unit back and forth across a transition edge, without rotation. This cannot happen with a quadrature encoder. Even if the encoder stopped on a transition and vibration caused the output to show transitions on one channel, there would be no transition on the other channel. By using quadrature detection on a two-channel encoder and viewing the transition in its relationship to the condition of the other channel, we can generate accurate and reliable position and direction information.

Pulses from the A and B channels can be fed to an up-down counter or programmable controller input card. Many PLC high-speed input modules have the capability of accepting encoder input. With quadrature detection, we can derive 1, 2, or 4 times the encoder's basic resolution. For example, if a 1000-count encoder is used, its basic resolution is 1000 pulses per revolution, or 360 degrees/1000. If the detection circuit looks at the high-to-low transition and the high-to-low transition of the B channel, we would have 2000 pulses per revolution. If the detection

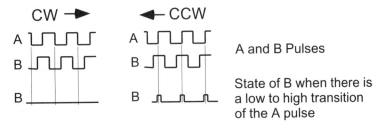

Figure 9–12 Determination of rotation direction. Encoder pulses for clockwise and counterclockwise rotation.

Figure 9–13 Disk from an absolute encoder. (*Courtesy BEI Sensors & Systems Company.*)

8 Bit Absolute Disc

circuit looks at the high-to-low transition and the low-to-high transition for the A and the B channels, we would have 4000 pulses per revolution. With a quality encoder, 4 times the signal will be accurate to better than 1/5 pulse.

Absolute Encoders The absolute encoder provides a whole word of output with a unique pattern that represents each position (see Figure 9-13). The LEDs and receivers are aligned to read the disk pattern (see Figure 9-14). Many types of coding schemes can be used for the disk pattern; the most common patterns are gray, natural, binary, and binary-coded decimal (BCD). Gray and natural binary are available up to about 256 counts (8 bits). Gray code is very popular because it is a nonambiguous code. Only one track changes at a time. This allows any inde-

Figure 9–14
Photodetectors read the position of an absolute encoder. (*Courtesy BEI Sensors & Systems Company.*)

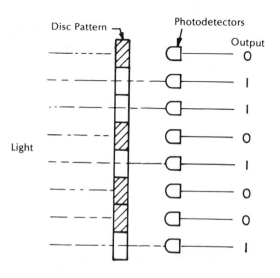

cision that may occur during any edge transition to be limited to plus or minus 1 count. Natural binary code can be converted from gray code through digital logic. If the output changes while it is being read, a latch option locks the code to prevent ambiguities.

Encoder Wiring Encoders are available in two wiring types: single ended and differential. Single-ended encoders have one wire for each ring (A, B, and X) and a common ground wire. A differential encoder has a separate ground for each ring and is more immune to noise because the rings do not share a common ground. The differential type is the most common in industrial applications.

Ultrasonic Sensor An ultrasonic sensor uses high-frequency sound to measure distance by sending out sound waves and measuring how long it takes them to return. The distance to the object is proportional to the length of time it takes to return. See Figure 9-15. An ultrasonic sensor can make very accurate measurements; their accuracy for objects as small as 1 millimeter can be plus or minus 0.2 millimeter. Some cameras use ultrasonic sensing to determine the distance to the object to be photographed.

An *interferometer sensor* sends out a sound wave that is reflected back to the sensor. See Figure 9-16. The transmitted wave interferes with the reflected wave. If the peaks of the two waves coincide, the amplitude is twice that of the transmitted wave. If the transmitted wave is 180 degrees out of phase with the reflected wave, the amplitude of the result is zero. Between these two extremes the amplitude is between zero and twice the amplitude, and the phase will be shifted between 0 and 180 degrees.

Interferometer sensors can detect distances to within a fraction of a wavelength. This is very fine resolution indeed because some light has wavelengths in the size range of .0005 millimeters.

Figure 9–15 An ultrasonic sensor. Note the use of the "gate" sensor to notify the PLC when a part is present. (*Courtesy of Omron Electronics.*)

Measuring height of different objects on a conveyor using external gate input to coordinate inspection

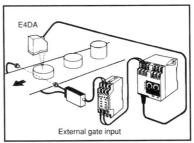

E4DA

External gate input

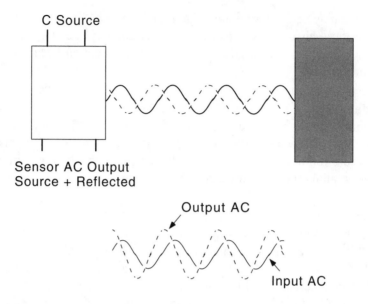

Figure 9–16 Ultrasonic sensor.

ELECTRONIC FIELD SENSORS

Electronic field sensors sense objects by producing an electromagnetic field. If the field is interrupted by an object, the sensor turns on. Field sensors are a better choice in dirty or wet environments than a photosensor, which can be affected by dirt, liquids, or airborne contamination.

The two most common types of field sensors, capacitive and inductive, function in essentially the same way. Each has a field generator and a sensor to sense when the field is interfered with. The field generator puts out a field similar to the magnetic field from a magnet.

Inductive Sensor

Used to sense metallic objects, the inductive sensor works by the principle of electromagnetic induction (see Figure 9-17). It functions in a manner similar to the primary and secondary windings of a transformer. The sensor has an oscillator and a coil; together they produce a weak magnetic field. As an object enters the sensing field, small eddy currents are induced on the surface of the object. Because of the interference with the magnetic field, energy is drawn from the sensor's oscillator circuit, decreasing the amplitude of the oscillation and causing a

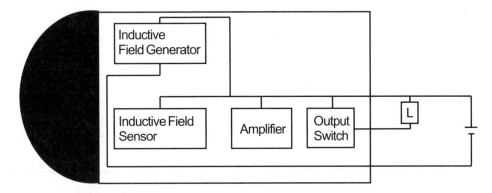

Figure 9–17 Block diagram of an inductive sensor. The inductive field generator creates an inductive field in front of the sensor; the field sensor monitors this field. When metal enters the field, the disruption in the field is sensed by the field sensor, and the sensor's output changes state. The sensing distance of these sensors is determined by the size of the field. This means that the larger is the required sensing range, the larger is the diameter of the sensor.

voltage drop. The sensor's detector circuit senses the voltage drop of the oscillator circuit and responds by changing the state of the sensor.

Sensing Distance Sensing distance (range) is related to the size of the inductor coil and whether the sensor coil is shielded or nonshielded (see Figure 9-18). This sensor coil is shielded in this case by a copper band. This prevents the field from extending beyond the sensor diameter and reduces the sensing distance. The shielded sensor has about half the sensing range of an unshielded sensor. Sensing distance generally varies by about 5 percent due to changes in ambient temperature.

Hysteresis Hysteresis means that an object must be closer to a sensor to turn it on than to turn it off (see Figure 9-19). Direction and distance are important. If the object is moving toward the sensor, it must move to the closer point to turn on. Once the sensor turns on (*operation point* or *on-point*), it remains on until the

Figure 9–18 A copper band in a shielded field sensor. Note that the sensing distance is reduced. If the unshielded sensor were mounted flush, it would detect the object in which it was mounted.

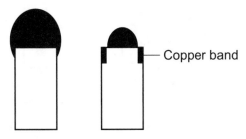

Copper band

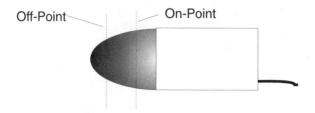

Figure 9–19   Example of hysteresis. Note that the on-point and off-point are different.

object moves to the *release point* (*off-point*). Hysteresis causes this differential gap, or differential travel. The principle is used to eliminate the possibility of "teasing" the sensor. The sensor is either on or off.

Hysteresis is a built-in feature in proximity sensors that helps stabilize part sensing. Imagine a bottle moving down a conveyor line. Vibration causes the bottle to wiggle as it moves along the conveyor. If the on-point was the same as the off-point and the bottle was wiggling as it went by the sensor, it could be sensed many times as it wiggles in and out past the on point. When hysteresis is involved, however, the on-point and off-point are at different distances from the sensor.

To turn the sensor on, the object must be closer than the on-point. The sensor output remains on until the object moves farther away than the off-point, preventing multiple unwanted reads.

Capacitive Sensors

Capacitive sensors (Figures 9-20 and 9-21) can sense both metallic and nonmetallic objects as well as product inside nonmetallic containers (see Figure 9-22). These sensors are commonly used in the food industry and to check fluid and solid levels inside tanks.

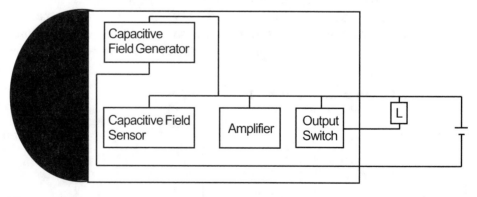

Figure 9–20   Block diagram of a capacitive sensor.

Figure 9–21 Capacitive sensors. (*Courtesy ifm efector inc.*)

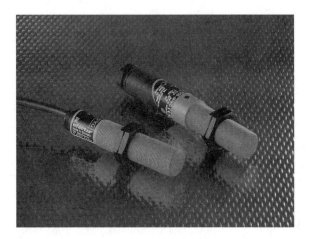

Capacitive sensors operate on the principle of electrostatic capacitance in a manner similar to the plates of a capacitor. The oscillator and electrode produce an electrostatic field (remember that the inductive sensor produced an electromagnetic field). The target (object to be sensed) acts as a capacitor's second plate. An electric field is produced between the target and the sensor. As the amplitude of the oscillation increases, the oscillator circuit voltage increases, and the detector circuit responds by changing the state of the sensor.

A capacitive sensor can sense almost any object. The object acts as a capacitor to ground. The entrance of the target (object) in the electrostatic field disturbs the DC balance of the sensor circuit, starting the electrode circuit oscillation and maintaining the oscillation as long as the target is within the field.

Sensing Distance Capacitive sensors are unshielded, nonembeddable devices (see Figure 9-21). This means that they cannot be installed flush in a mount because they would then sense it. Conducting materials can be sensed farther away than nonconductors because the electrons in conductors are freer to move. The target mass affects the sensing distance: The larger the mass, the larger the sensing distance.

Figure 9–22 Capacitive sensor checking inside boxes. (*Courtesy ifm efector inc.*)

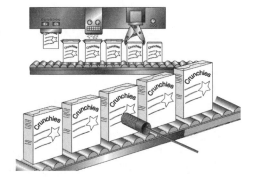

Capacitive sensors are more sensitive to temperature and humidity fluctuation than are inductive sensors, but capacitive sensors are not as accurate as inductive ones. Repeat accuracy can vary by 10 to 15 percent in capacitive sensors. Sensing distance for capacitive sensors can fluctuate as much as plus or minus 15 to 20 percent.

Some capacitive sensors are available with an adjustment screw, which can be adjusted to sense a product inside a container (see Figure 9-22). The sensitivity can be reduced so that the container is not sensed but the product inside is.

SENSOR WIRING

Both AC- and DC-powered sensors have basically two wiring schemes: *load powered* (two wire) and *line-powered* (three wire). The most important power consideration is to limit the sensor's output current to an acceptable level. If the output current exceeds the output current limit, the sensor will fail. The sensor's output device is normally a transistor. A sensor with a transistor output can generally handle up to about 100 milliamps. Some sensors have a relay output and can handle 1 amp or more.

Load-Powered Sensors

In two-wire or load-powered sensors, one wire is connected to power, and the other wire is connected to one of the load's wires (see Figure 9-23). The load wire is then connected to power. Wiring diagrams can usually be found on the sensor.

The current for the sensor is typically under 2.0 milliamps (see Figure 9-24), which is enough for the sensor to operate but not enough to turn on the PLC's input. The current required for the sensor to operate must pass through a *load*, which is anything that will limit the current of the sensor output. Think of the load as an input to a PLC. The small current that must flow to allow the sensor to operate is called *leakage current* or *operating current*. (The leakage current is usually not enough current to activate a PLC input. If it is enough, it must be con-

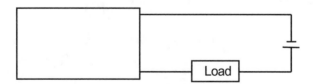

Figure 9–23 Two-wire (load-powered) sensor. The load represents whatever the technician monitors the sensor output for, normally a PLC input. The load must limit the sensor's current to an acceptable level, or the sensor output blows.

Figure 9–24 Leakage current vs. supply voltage.

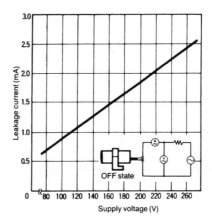

nected to a bleeder resistor as shown in Figure 9-25.) When the sensor turns on, it allows enough current to flow to turn on the PLC input.

Response time, which can be crucial in high-demand production applications, is the lapsed time between sensing the target and changing the output state. Sensor specification sheets provide response time information.

Line-Powered Sensors

Although line-powered sensors are usually of the three-wire type (see Figure 9-26), they can have either three or four wires. The three-wire variety has two power leads and one output lead.

The line-powered sensor needs a small current, called *burden current* or *operating current*, to operate. This current flows whether the sensor output is on or off. The load current is the output from the sensor; if it is on, there is load current. This load current turns on the load (PLC input). The maximum load current is typically between 50 to 200 milliamps for most sensors. The output LED on a sensor can function even though the output is bad.

Figure 9–25 Use of a two-wire (load-powered) sensor. In this case, the leakage current was large enough to cause the input module to sense an input when there was none. A resistor was added to bleed the leakage current to ground so that the input could not sense it.

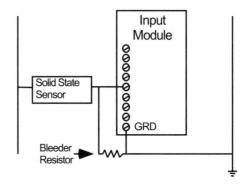

Figure 9–26 Three-wire (line-powered) sensor. The load must limit the current to an acceptable level.

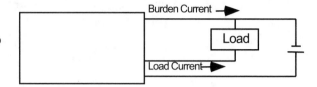

Sourcing and Sinking Sensors

Sourcing Sensor (PNP)

In the case of a sourcing sensor (PNP), when there is an output current from the sensor, the sensor sources current to the load (see Figure 9-27). Conventional current flow goes from plus to minus. When the sensor is off, the current does not flow through the load.

Sinking Sensor (NPN)

In the case of sinking (NPN) sensors, when the sensor is off (nonconducting), no current flows through the load. When the sensor is on (conducting), a load current flows from the load to the sensor (see Figure 9-28). The choice of whether to use an NPN or a PNP sensor depends on the type of load. In other words, choose a sensor that matches the PLC input module requirements for sinking or sourcing.

Sinking = Path to supply ground (–)

Sourcing = Path to supply source (+)

Analog Sensors

Many types of analog sensors are available. Many of the types already discussed are available with digital and analog output. Photosensors and field sensors are available with analog output.

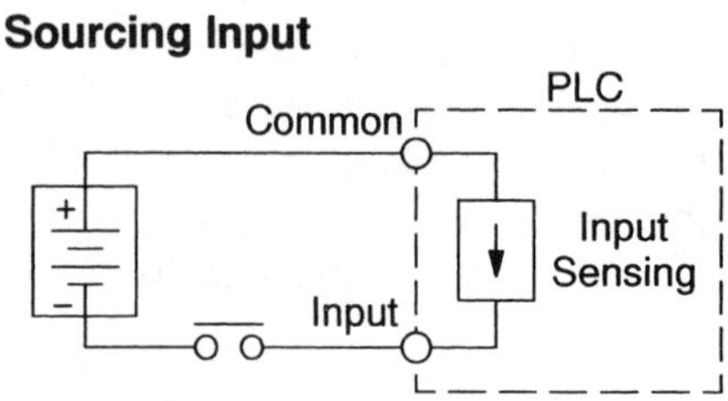

Figure 9–27 A sourcing switch connected to a PLC input. (*Courtesy PLC Direct by Koyo.*)

Sourcing Input

Figure 9–28 A sinking switch connected to a PLC input. *(Courtesy PLC Direct by Koyo.)*

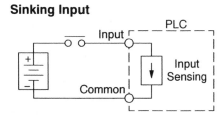

Sinking Input

Analog Considerations

Analog sensors provide much more information about a process than digital sensors do. Their output varies depending on the conditions being measured.

Accuracy, Precision, and Repeatability

Accuracy can be defined as how closely a sensor indicates the true quantity being measured. In a temperature measurement, accuracy is defined as how closely the sensor indicates the actual temperature being measured. *Precision* refers to how closely each sensor in a group of sensors measures the same variable. For example, when measuring a given temperature, the output of each should be the same. Precision is very important in many applications. *Repeatability* is a sensor's ability to repeat its previous readings. For example, in measuring temperature, the sensor's output should be the same every time it senses a specific temperature.

Thermocouples

The principle of thermocouple was discovered by Thomas J. Seebeck in 1821. The *thermocouple* is one of the most common devices for temperature measurement in industrial applications. A *thermocouple* is a very simple device that has two pieces of dissimilar metal wire joined at one or both ends. The typical industrial thermocouple is joined at one end (see Figure 9-29). The other ends of the wire are connected via compensating wire to the analog inputs of a control device such as a PLC (see Figure 9-30). The operation principle is that joining dissimilar metals produces a small voltage. The voltage output is proportional to the difference in temperature between the cold and hot junctions. Thermocouples are colorcoded for polarity and for type. The negative terminal is red, and the positive terminal is a different color that can be used to identify the thermocouple type (see Figure 9-31).

A cold junction is assumed to be at ambient temperature (room temperature). In reality, temperatures vary considerably in an industrial environment. If the cold junction varies with the ambient temperature, the readings are inaccurate,

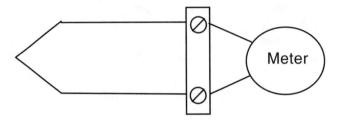

Figure 9–29 Simple thermocouple. A thermocouple can be made by twisting the desired type of wire together and silver soldering the end. To measure the temperature change, the wire is cut in the middle to insert a meter. The voltage is proportional to the difference in temperature between the hot and cold junctions.

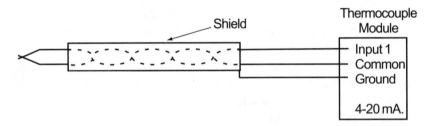

Figure 9–30 Thermocouple. The wire connecting the thermocouple to the PLC module is twisted-pair (two wires twisted around each other) shielded cable. The shield around the twisted pair helps eliminate electrical noise as a problem. Twisted-pair wiring also helps reduce the effects of electrical noise. The shielding is grounded only at the control device. Typical PLC modules allow four or more analog inputs.

Type	+IMeta	- Meta	Average Microvolt Change Per Degree F.		Useful Range in Degrees Centigrade
J)	Iron (White)	Constantan (Red)	2	29.	0 0-80
K)	Chromel (Yellow)	Alumel (Red)	5	21.	0 0-120
R)	Platinum (Black)	Rhodium (Red)	7	6.	500-150
T)	Copper (Blue)	Constantan (Red)	5	22.	0-100-40

Figure 9–31 Thermocouple color table.

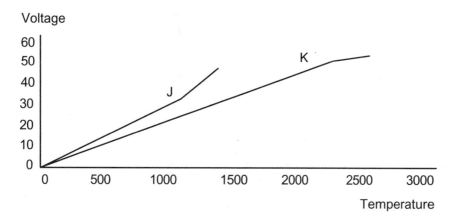

Figure 9–32 Voltage output vs. temperature for J- and K-type thermocouples. The relationship is approximately linear. For example, if the output voltage was 20 millivolts with a J-type thermocouple the temperature would be approximately 525 degrees (600 − 75). Note the voltage produced by the difference in temperature between the cold and hot junctions. The chart assumes a 75 degree ambient temperature.

which is unacceptable in most industrial applications. Industrial thermocouple tables use 75 degrees Fahrenheit for the reference temperature (see Figure 9-32), but it is too complicated to try to maintain the cold junction at 75 degrees, so industrial thermocouples must be compensated. This is normally accomplished with the use of temperature-sensitive resistor networks. The resistor used in the network has a negative coefficient of resistance. Because resistance decreases as the temperature increases, the voltage automatically adjusts so that readings remain accurate. PLC thermocouple modules automatically compensate for temperature variation.

The resolution of a thermocouple, which provides accurate measurements, is determined by the device that takes the output from the thermocouple. The device is normally a PLC analog module. The typical resolution of an industrial analog module is 12 bits; 2 to the twelfth power is 4096. This means that if the range of temperature to be measured was 1200 degrees, the resolution would be 0.29296875 degrees/bit (1200/4096 = 0.29296875), which means that the PLC could tell the temperature to about one-fourth of one degree. This is reasonably good resolution, close enough for the vast majority of applications, and higher-resolution analog modules are available for higher requirements. See Figure 9-33 for the composition of some of the various thermocouples available. Figure 9-34 shows a comparison of the voltage output versus temperature for a J-type and K-type thermocouple. Figure 9-34 shows several types of thermocouples and

Thermocouple Types
Nickel-Chromium vs. Copper-Nickel (Chromel-Constantan)
Iron vs. Copper-Nickel (Iron-Constantan)
Nickel-Chromium vs. Nickel-Aluminum (Chromel-Alumel)
Copper vs. Copper-Nickel (Copper-Constantan)
Tungsten 5% Rhenium vs. Tungsten 26% Rhenium
Tungsten vs. Tungsten 26% Rhenium
Tungsten 3% Rhenium vs. Tungsten 25% Rhenium
Platinum vs. Platinum 13% Rhodium
Platinum vs. Platinum 10% Rhodium
Platinum 6% Rhodium vs. Platinum 30% Rhodium

Figure 9–33 Composition of various thermocouple types.

their temperature ranges. The user chooses the correct thermocouple for the range of temperature in the application.

Resistance Temperature Device (RTDs)

A resistance temperature device (RTD) is a precision resistor whose resistance changes with temperature. RTDs are more accurate than thermocouples. The most common RTD resistance is 100 ohms at zero degrees Celsius; others are available in the 50 to 100 ohm range. One of the basic properties of metals is that their electrical resistivity changes with temperature. Some metals have a very predictable change in resistance for a given change in temperature. The chosen metal is made into a resistor with a nominal resistance at a given temperature. The change in temperature can then be determined by comparing the resistance at the unknown temperature to the known nominal resistance at the known temperature. Tables of the temperature-resistance relationships for various metals used in RTDs are available.

RTDs are made from a pure wire-wound metal, usually one that occurs naturally, that has a positive temperature coefficient. As the temperature increases, the very small change in resistance to current flow increases.

RTDs come in two configurations, a wire-wound construction and a thin film metal layer construction deposited on a nonthermally conductive substrate. The

Thermocouple Temperature Ranges

Type	Temperature Fahrenheit	Temperature Celsius
J	−50 to +1400	−40 to +760
K	−100 to +2250	−80 to +1240
T	−200 to +660	−130 to +350
E	−200 to +1250	−130 to +680
R	0 to +3200	−20 to +1760
S	0 to +3200	−20 to +1760
B	+500 to +3200	+260 to +1760
C	0 to +4200	−20 to +2320

Figure 9–34 Temperature ranges for various types of thermocouples. Note the output for the difference in temperature between the hot and cold junctions. The table is based on a cold junction temperature of 75 degrees Fahrenheit.

wire-wound coiled construction allows for a greater change in resistance for a given change in temperature, increasing the sensor's sensitivity and improving the resolution. The coil is wrapped around a nonconductive material such as ceramic, which enables the wire to respond more quickly to temperature change. The coil is then covered by a sheath made from very temperature-conductive materials to protect it from abuse and the environment. The sheath is filled with a dry thermal-conducting gas to increase the temperature conductivity.

Platinum is the most popular material for RTDs. It has a very linear change in resistance versus temperature and has a wide operating range. Platinum is a very stable element, which ensures long-term stability. Platinum sensors are now being made with very thin film resistance elements that use only a very small amount of platinum, which makes the platinum RTDs more competitive in price. Other materials include copper, nickel, balco, tungsten, and iridium.

The second RTD configuration is in the form of a thin film metal layer deposited on a nonthermally conductive substrate, which is often ceramic. These RTDs can be very small and do not require an external sheath because the ceramic acts as a protective sheath. Simple circuitry can be used to linearize RTDs. RTDs are more accurate than thermocouples.

Wiring RTD can have three different wiring configurations (see Figure 9-35). The lead wires from the RTD can affect its accuracy because it is additional uncompensated resistance. A third wire can be added to compensate for lead wire

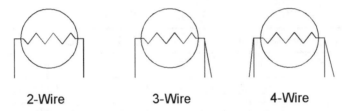

2-Wire   3-Wire   4-Wire

Figure 9–35   Three RTD wiring configurations.

resistance. Four-wire RTDs are also available but are generally used only in laboratory applications. The four-wire RTD can be used with a three-wire device but this does not improve a three-wire RTD's accuracy. A three-wire RTD cannot be used with a four-wire device. It is best to use the appropriate RTD for the measurement device.

Thermistors

A *thermistor* is a temperature-measuring sensor constructed of man-made materials. It is more sensitive to temperature than an RTD. A thermistor has a negative temperature coefficient. The resistance decreases as the temperature increases. Since a thermistor is a semiconductor, it cannot operate above about 300 degrees Celsius. Major advantages of a thermistor are precision, stability, and production of a large change in resistance for a small change in temperature. If the range of temperature to be measured is relatively small, the thermistor is a good device. One of the thermistor's main problems is that its output is linear only within a small temperature range (see Figure 9-36). Its resistance does not vary proportionately with a change in temperature, although thermistor networks have very linear voltage change with temperature change.

Different package styles of thermistors are available for various applications, such as monitoring the temperature of electric motors. The thermistor is fastened to the housing of the motor and connected to a bridge circuit whose output is compared to a reference voltage. The reference voltage is chosen to be a safe value for the motor's maximum operating temperature. It can then be used to shut off the motor circuit.

Integrated Circuit (Semiconductor) Temperature Sensors

Some integrated circuit semiconductors can be used to sense temperature change. They respond to temperature increases by increasing their reverse bias current across PN junctions and generating a small, detectable current or voltage that is proportional to the change in temperature (see Figure 9-37).

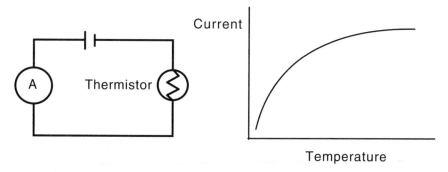

Figure 9–36 Graph of current versus temperature change for a thermistor.

Magnetic Reed Sensors

Magnetic reed sensors usually have two sets of contacts, a normally closed set and a normally open set. When a small magnet is brought close to the reed sensor, the common contact is drawn toward the normally open contact and moves away from the normally closed contact. These magnetic reed sensors are commonly used for sensing whether a piston in a cylinder is extended or retracted for feedback in a system.

Hall Effect Sensor

The hall effect sensor is based on a principle discovered shortly after the transistor was developed in the 1950s. A hall effect sensor is biased with a current. If a magnetic field is introduced to the sensor material, it allows a small current to flow through the sensor and then can be sensed (see Figure 9-38). Hall effect sensors are available as both digital and analog.

Hall effect sensors have many applications. They are used in cruise (speed) control on cars, as blood flow sensors in the medical field, as linear displacement

Figure 9–37 Graph of voltage versus temperature change for an integrated circuit (semiconductor).

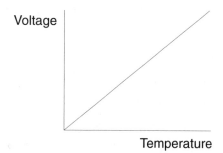

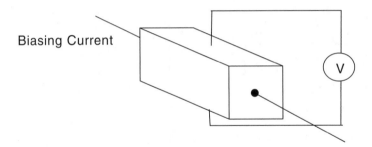

Figure 9–38   Basic principle of the hall effect sensor.

devices and as sensors of rotary motion for speed. They are often used to detect the velocity of a rotating shaft and to determine displacement of a shaft.

Strain Gages

Strain gages measure force based on the principle that the thinner the wire is, the higher the resistance is. A wire with a smaller diameter would have higher resistance than a wire with a larger diameter (see Figure 9-39). If we had an elastic wire, we could stretch it and measure its change in resistance because its diameter decreases in the middle, increasing the resistance. The change in resistance compared to the change in voltage is quite linear. If a constant current is supplied to a strain gage and the force applied to the strain gage varies, the resistance of the strain gage will vary and the voltage change will be proportional to the change in force.

Strain gages have a number of uses. They are used for pressure measurement by bonding them to a membrane that is then exposed to the pressure and for weight measurement. They are also used in accelerometers. Most strain gages are zig-zag wire mounted on a paper or membrane backing (see Figure 9-40). They must be mounted correctly because they are sensitive to change in only one direction. Strain gages normally have an arrow that indicates the direction in which they should be mounted. They are normally mounted with adhesives.

They are often set up in a bridge configuration to increase the change in resistance and to compensate for temperature fluctuations (see Figure 9-41). To

Figure 9–39   Two pieces of wire. Which one will have lower resistance? The larger the wire diameter, the lower is the resistance.

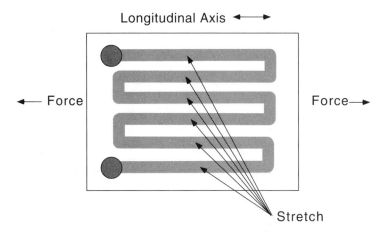

Figure 9–40 Strain gage zig-zag mounted on paper or membrane backing.

do this, a dummy strain gage is inserted and is not subjected to the strain but to the same temperature change. This eliminates the effect of temperature fluctuation in the stress measurement. Since strain gages are made from long metal strands (or metallic paths) and metal responds to temperature change, strain gages are susceptible to variation in temperature, which can cause a variance in the output.

Strain gages can be set up in another way to overcome fluctuation due to variation in temperature (see Figure 9-42). In this case four strain gages and two balancing resistors (R5 and R6) are used to compensate for thermal drift and to balance the bridge. All of the strain gages will adapt to any temperature fluctuation and will thus cancel any temperature-induced fluctuation.

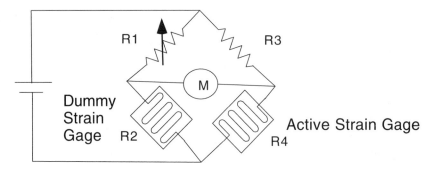

Figure 9–41 Bridge configuration of a strain gage to compensate for temperature fluctuations.

Figure 9–42 Strain gage set up to compensate for temperature fluctuation.

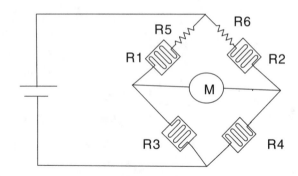

Linear Variable Displacement Transformer (LVDT)

A linear variable displacement transformer (LVDT) uses the transformer principle to sense position. It is a cylindrical sensor with three fixed windings. The cylinder's center is hollow, which allows a core to slide in the cylinder (see Figure 9-43). The primary coil is connected to a low-voltage AC source (a sine wave). As the core moves through the LVDT, a voltage is induced in the other two windings. The voltage is proportional to the core position.

   LVDTs can be wired in two ways: the series-adding mode or the series opposition mode. In the series-adding mode, the secondary coils are wired in series, adding their output. In the series opposition the leads from one secondary are reversed and then the secondaries are wired in series. The output voltages are then in opposition. The series-adding mode increases the voltage smoothly as the core moves into and through the core (see Figures 9-44 and 9-45). The voltage in the series opposition mode is zero when the core is centered and increases as the core moves from the centered position (see Figure 9-46). The signal also changes phase on each side of the centered position. Figures 9-47, 9-48, and 9-49 show examples of the output for an LVDT that is wired in the series opposition mode. The output from an LVDT can be rectified to provide a DC signal (see Figure 9-50) because it is easier for PLCs to work with a DC signal than an AC signal. This is called signal conditioning. LVDTs can be purchased with DC output.

Figure 9–43 Typical LVDT. The core is aligned with both windings. Each of the output windings is 6 VAC with a phase of 180 degrees.

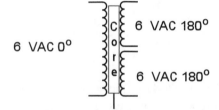

Figure 9–44 Series-adding method of use with output of 12 VAC 180 degrees added.

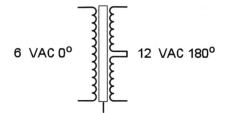

6 VAC 0°   12 VAC 180°

Figure 9–45 Total output in the series-adding mode whose core is located halfway into the windings is 6 VAC 180 degrees.

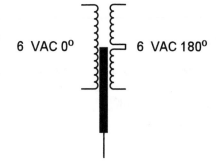

6 VAC 0°   6 VAC 180°

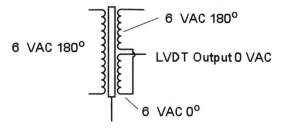

6 VAC 180°

6 VAC 180°

LVDT Output 0 VAC

6 VAC 0°

Figure 9–46 Series opposition method with the core centered. The output of the top winding is 6 VAC 180 degrees. The bottom winding is 6 VAC 0 degrees because, in effect, the leads have been reversed. The output when the core is centered is 0 volts.

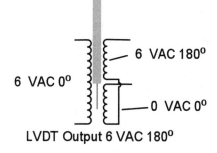

6 VAC 180°

6 VAC 0°

0 VAC 0°

Figure 9–47 LVDT core is aligned with the top output winding. The output is 6 volts 180 degrees.

LVDT Output 6 VAC 180°

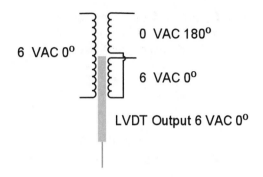

Figure 9–48   LVDT whose core is aligned with the lower winding. The output is 6 VAC 0 degrees.

Resolvers

A resolver is an analog position transducer that uses the principle of the transformer to operate. It is a very reliable device. The bearings are really the only element that can fail.

In a resolver, a 0-degree sine wave (AC voltage) is applied to one stationary coil (see Figure 9-51). A 90-degree sine (cosine) wave is applied to a second stationary winding. The frequency of these signals is typically about 2 kilohertz. The resolver's rotor has a third winding that rotates with the rotor. Because of the transformer action, this winding produces an output. Usually the rotor has half as many windings as there are on the stator windings, so the output voltage will be one-half the input voltage. If the rotor is aligned with the 0-degree sine winding, the output is 6 volts 180 degrees out of phase (remember the transformer). So if we put the resolver rotor winding output on a scope, it would be 6 volts 180 degrees out of phase. See Figure 9-52. If the rotor is rotated one-half turn (180 degrees), it aligns with the 0-degree sine winding again, but the output is 6 volts 0 degrees (see Figure 9-52).

When the rotor is aligned with the 90-degree stator, the output is 6 volts 270 degrees; the output is 180 degrees out of phase with the input 90-degree sine wave (see Figure 9-53). When the rotor is rotated one-half turn (180 degrees), the output is 6 volts 90 degrees (see Figure 9-54).

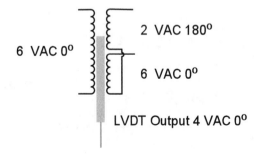

Figure 9–49   Core aligned with the lower winding and part of the upper winding. The output of the LVDT in the series opposition mode is 4 VAC 0 degrees.

Figure 9–50 Output from an LVDT can be rectified to provide a DC signal that can be read by an analog input card.

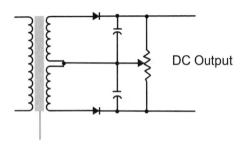

DC Output

Figure 9–51 The rotor is aligned with stator winding 1. The rotor output is 6 VAC (one-half the number of windings on the rotor) 180 degrees.

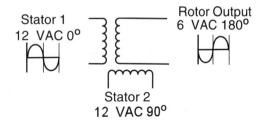

Stator 1
12 VAC 0°

Rotor Output
6 VAC 180°

Stator 2
12 VAC 90°

Figure 9–52 Rotor is aligned with the 0-degree stator but has been rotated 180 degrees.

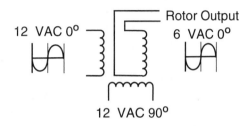

12 VAC 0°

Rotor Output
6 VAC 0°

12 VAC 90°

Figure 9–53 Rotor lined up with the 90-degree stator.

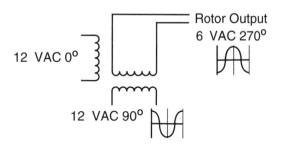

Rotor Output
6 VAC 270°

12 VAC 0°

12 VAC 90°

Figure 9–54 Output with the rotor aligned with the 90-degree stator but revolved 180 degrees.

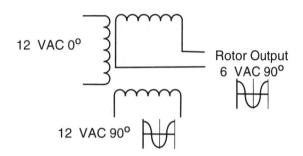

12 VAC 0°

Rotor Output
6 VAC 90°

12 VAC 90°

185

Figure 9–55 Output if the rotor were at a 45-degree angle to the 0- and 90-degree stators.

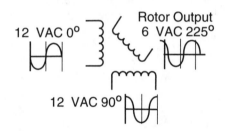

What if the rotor is not aligned with either stator but is located at some other angle? Examine Figure 9-55. In this case, the rotor is at a 45-degree angle to the 0- and 90-degree stators. In this case the rotor winding output is a combination of the field created by the 0- and 90-degree stator. The output is 6 volts 225 degrees. If the rotor is now turned one-half turn (180 degrees), the output is 6 volts 45 degrees. The rotor output is always 6 volts. The phase of the rotor output is determined by the orientation of the rotor in relation to the stator windings. A computer is capable of comparing the 0-degree sine wave to the output phase. By comparing them, the computer always knows the position of the rotor. One manufacturer of position controllers uses a clock that pulses 4000 times per cycle to compare the phase of the input and output and to break the cycle into 4000 pieces. This gives a resolution of 4000 for 360 degrees of rotation, which is very fine.

Resolvers also are used in the ratiometric tracking method, which is the reverse of the type just studied. The ratiometric tracking method applies a sine wave input to the rotor. This causes an output voltage and phase on the stators. As the rotor turns, the stator voltage and phase vary. The computer can compare the two stator outputs and determine position (see Figure 9-56).

Pressure Sensors

Pressure sensors typically measure and control fluids such as gases and liquids. Some pressure sensors operate through a change in resistance, some through a change in capacitance, and some through changes in inductance. The strain gage pressure sensor discussed earlier has attached a strain gage to a membrane that stretches in proportion to the pressure applied to it (see Figure 9-57). If a constant current is applied to the strain gage, its output voltage changes corresponding to the change in pressure.

Another pressure sensor utilizes an LVDT (see Figure 9-58). Pressure applied to the bellows causes it to expand and move the core in the LVDT. This change in core position can then be measured by the output of the LVDT.

Industrial networks (buses) can minimize installation and maintenance costs of sensors. Figure 9-59 shows examples of some industrial sensors and networks.

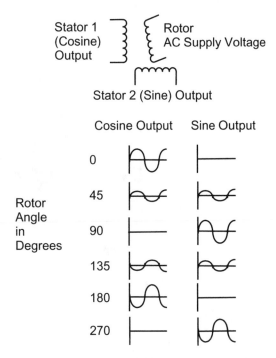

Figure 9–56 Output for a resolver using the ratiometric method.

Stator 1 (Cosine) Output

Rotor AC Supply Voltage

Stator 2 (Sine) Output

Cosine Output   Sine Output

Rotor Angle in Degrees

0

45

90

135

180

270

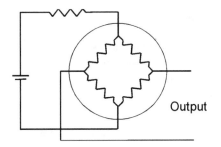

Figure 9–57 Strain gage pressure sensor. The circle represents a membrane to which the pressure is applied.

Output

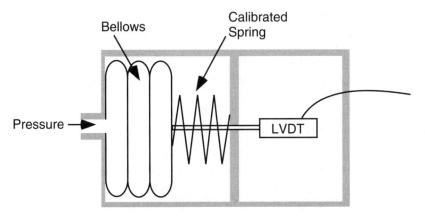

Bellows

Calibrated Spring

Pressure →

LVDT

Figure 9–58 Bellows-style pressure sensor. Any change in pressure affects the bellows and moves the core in the LVDT.

Figure 9–59   Industrial sensors and networks. (*Courtesy ifm efector inc.*)

SENSOR INSTALLATION CONSIDERATIONS

Electrical

The main installation consideration for sensors is to limit the load current. The output (load) current must be limited on most sensors to a very small output current. The output limit is typically between 50 and 200 milliamps. If the load draws more than the sensor current limit, the sensor fails and must be replaced. More sensors probably fail because of improper wiring than from use. It is crucial that current be limited to a level that the sensor can handle. PLC input modules limit the current to acceptable levels, and sensors with relay outputs can handle higher load currents (typically 3 amps).

If high-voltage wiring is run in close proximity to sensor cable, the cable should be run through a metal conduit to prevent the sensor from false sensing, malfunction, or damage. The other main consideration is to buy the proper polarity sensors. If the PLC module requires sinking devices, they should be installed.

Mechanical

Mechanical sensors should be mounted horizontally whenever possible to prevent the buildup of chips and debris on the sensor that could cause misreads. In a vertical position, chips, dirt, oil, and so on can gather on the sensing surface and cause the sensor to malfunction. In a horizontal position, the chips fall away. If the sensor must be mounted vertically, provision must be made to remove chips and dirt periodically using air blasts or oil baths.

Care should be taken so that the sensor does not detect its own mount. For example, an inductive sensor mounted improperly in a steel fixture might sense the fixture. If two sensors are mounted too close together, they can interfere with each other and cause erratic sensing.

TYPICAL APPLICATIONS

One of the most common uses of a sensor is in product feeding when parts move along a conveyor or in some type of parts feeder. The sensor notifies the PLC when a part is in position and is ready to be used. This is typically called a *presence/absence check*. The same sensor can also provide the PLC with additional information that the PLC uses to count the parts as they are sensed. The PLC can also compare the completed parts and elapsed time to compute cycle times and can track scrap rates, production rates, and cycle time.

One simple sensor allowed the PLC to accomplish three tasks:

Are there parts present?

How many parts have been used?

What is the cycle time for each part?

Simple sensors can be used to decide which product is present. Imagine a manufacturer that produces three different package sizes on the same line. The product sizes move randomly along a conveyor. When each package arrives at the end of the line, the PLC must know which size of product is present. This can be done very easily with three simple sensors. If only one sensor is on, the small product is present. If two sensors are on, the middle-size product is present. If three sensors are on, the product is the large size. The same information could then be used to track production for all product sizes and cycle times for each size.

Sensors can be used to check whether containers have been filled. Imagine aspirin bottles moving along a conveyor with the protective foil and the cover on. Simple sensors can sense right through the cap and seal and make sure that the bottle was filled. One sensor, often called a *gate sensor*, senses when a bottle is present. A gate sensor shows when a product is in place. The PLC then knows that a product is present and can perform other checks.

A second sensor senses the aspirin in the bottle. If a bottle is present, but the sensor does not detect the aspirin inside, the PLC knows that the aspirin bottle was not filled.

As discussed earlier, sensors can monitor temperature. Imagine for example, a sensor can monitor the temperature in an oven in a bakery. The PLC can then control the heater element in the oven to maintain the ideal temperature.

Pressure is vital in many processes. Plastic injection machines force heated plastic into a mold under a given pressure. Sensors can monitor the pressure, which must be accurately maintained or the parts will be defective. The PLC can monitor the sensor and control the pressure.

Flow rates are important in process industries such as papermaking. Sensors can monitor the flow rates of fluids and other raw materials. The PLC can use these data to adjust and control the system's flow rate. The water department monitors the flow rate of water to calculate bills.

Figure 9-60 is an illustration of a temperature and flow application. The flow sensor on the upper left of the machine monitors the proper flow of cooling water into the chiller. A built-in display displays the flow setpoint and the flow status. The sensor on the lower right of the machine is a temperature sensor that can be set to the proper temperature and alarm point.

The applications noted here are a few of the simple uses for sensors. Innovative engineers or technicians invent many other uses. The data from one sensor can be used to provide many different types of information (e.g., presence/absence, part count, cycle time).

When choosing the sensor to use for a particular application, several important considerations, such as the material in the object to be sensed, are crucial. Is the material plastic? Is it metallic? Is it a ferrous metal? Is the object clear, transparent, or reflective? Is it large or very small?

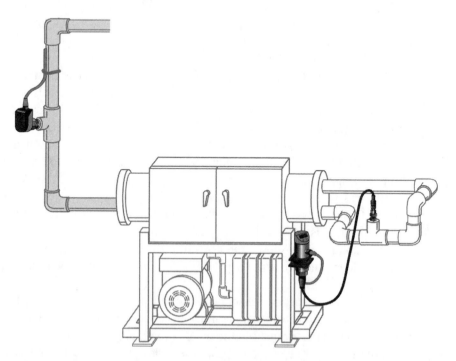

Figure 9–60   Temperature and flow application. (*Courtesy ifm efector inc.*)

Figure 9–61 Sensor that senses empty and full levels in a container. (*Courtesy ifm efector inc.*)

The specifics of the physical application determine the sensor type to use. Is a large area available on which to mount the sensor? Are contaminants a problem? What speed of response is required? What sensing distance is required? Is excessive electrical noise present? What accuracy is required? Answering these questions helps narrow the choice, which must be made based on criteria such as the sensor's cost and reliability as well as the cost of failure. The cost of failure is usually the guide to how much sensing must be done. If the cost is high, sensors should be used to notify the PLC of problems.

Figure 9-61 shows a smart-level sensor. The sensor's microprocessor and push button are used to teach the PLC to recognize the container's empty and full conditions.

Two typical applications utilize ultrasonic sensors to measure distances. The application on the left uses the sensor's analog output to control the web precisely. The application on the right is measuring the height of objects. The fiber optic sensor is used as a gate to indicate part presence. The ultrasonic sensor then takes a reading and the height of the part is known. See Figure 9-62.

Numerous applications are shown in the following figures.

- Figure 9-63 shows a package sorting application that uses color mark sensors.
- Figure 9-64 is a photosensor used in a bottle-capping application. This sensor's output could be used to distinguish between bottles that are capped and those that are not. Note that a field sensor can work in hostile environments.
- Figure 9-65 shows the use of a laser sensor with a 5-micron resolution. The sensor uses a laser LED to emit a beam 10 millimeters wide with a maximum sensing distance of 300 millimeters (11.8 inches). In this case the analog output is used to detect wire breakage.

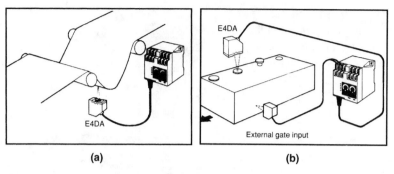

(a)   **(b)**

Figure 9–62 Two typical applications that use ultrasonic sensors to measure distances. (*Application concept courtesy of Omron Electronics.*)

Figure 9–63 Sorting packages using color mark sensors. (*Application concept courtesy of Omron Electronics.*)

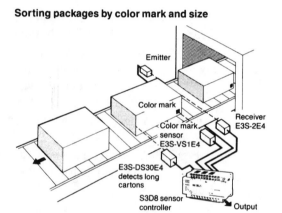

Figure 9–64 A photo sensor used to sense the proper capping of a container. (*Application concept courtesy of Omron Electronics.*)

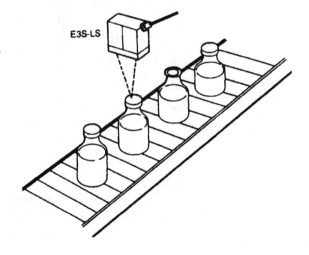

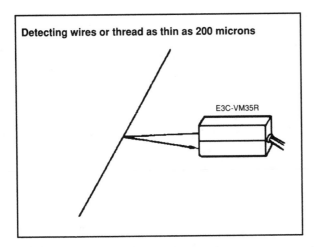

Detecting wires or thread as thin as 200 microns

E3C-VM35R

Figure 9–65 Wire can be sensed by a laser sensor. (*Application concept courtesy Omron Electronics.*)

- Figure 9-66 shows three fiber-optic applications used to sense small objects. In the first, the sensor checks for part presence. The second uses a special fiber-optic head to spread the beam. The third checks the diameter in a tape-winding application.

- In Figure 9-67 the sensor is mounted perpendicular to the transparent film to maximize the reflected light to the receiver as it detects the film.

- Figure 9-68 shows an inductive sensor being used to check to see the correct assembly of screws. This inductive sensor provides digital outputs for high/pass/low detection and could be used to pass or fail the parts into pass bins, too tall bins, and too short bins.

- Figure 9-69 is an example of a hydraulic pressure and level application. The pressure sensor on the upper left of the hydraulic power unit monitors the hydraulic pressure. The sensor on the lower right detects the level of hy-

Figure 9–66 Small objects can be sensed by a laser sensor. (*Application concept courtesy of Omron Electronics.*)

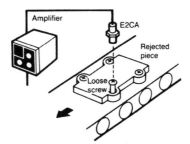

Amplifier

E2CA

Rejected piece

Loose screw

Figure 9–67 Detection of transparent film. (*Application concept courtesy of Omron Electronics.*)

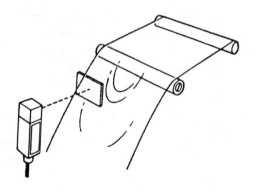

Detecting presence of printed circuit board components	Detects metal or non-metal chips within a sensing area as large as 2 x 11-mm	Detects when tape roll has reached selected diameter
	Chips need not be in a straight line. Sensor, used vertically, will detect object as tiny as 1.3 mm.	Outputs a signal when selected diameter is reached.

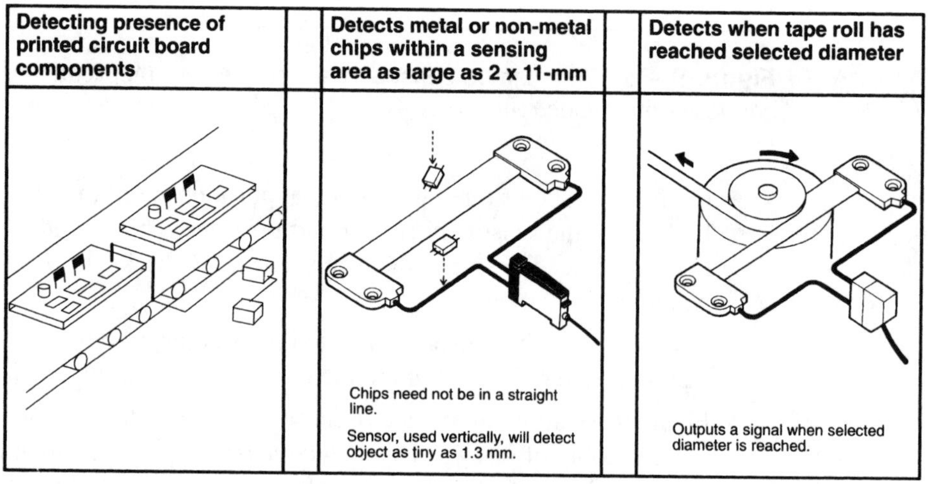

Figure 9–68 Detection of small metal objects. (*Application concept courtesy of Omron Electronics.*)

Figure 9–69 Hydraulic application. (*Courtesy ifm efector inc.*)

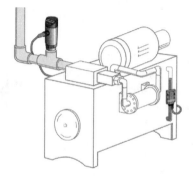

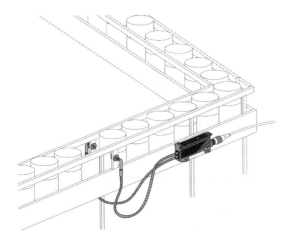

Figure 9–70   Fiber optic application in confined space. (*Courtesy ifm efector inc.*)

draulic fluid in the power unit. The sensor is mounted to a sight glass with a strap but it ignores the sight glass and senses only the liquid.

- Figure 9-70 is an application utilizing a fiber optic sensor in a confined space.
- Figure 9-71 shows capacitive sensors checking a box to determine that all nine bottles have been loaded into the carton.
- Figure 9-72 is a laser sensor used to check very small glass parts. The laser provides a very small diameter coherent beam to sense even very small parts.

Figure 9–71   Capacitive sensor application. (*Courtesy ifm efector inc.*)

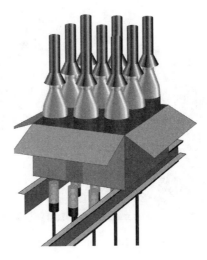

Figure 9–72   Laser sensor application of small size objects. The part is spherical, not flat. Only a very small area can reflect light back to the sensor. (*Courtesy ifm efector inc.*)

- Figure 9-73 shows a pressure sensor application that provides feedback to a controller for safety and proper operation. Digital and analog pressure sensors are available for this type of application.
- Figure 9-74 is an inductive sensor used to sense gear teeth as the gear rotates. The user could tell the position by the number of gear teeth that pass

Figure 9–73   Pressure-sensing application. (*Courtesy ifm efector inc.*)

Figure 9–74 Inductive sensor used to sense each tooth as it passes. (*Courtesy ifm efector inc.*)

or the velocity by the number of teeth that pass the inductive sensor in a given amount of time.

Choosing a Sensor for a Special Application

Applications always present special problems for the technician or engineer. The sensor may fail to sense every part or fail randomly. In this case, a different type or different model of sensor may be needed. Sales representatives and applications engineers from sensor manufacturers can be very helpful in choosing a sensor to meet a specific need. They have usually seen the particular problem before and know how to solve it. Magazines devoted to the topic of sensors are good sources of information.

The number of sensor types and the complexity of their use in solving application problems increases daily. New sensors are constantly being introduced to solve needs. Almost as many sensor types are available now as there are applications.

The innovative use of sensors can help increase the safety, reliability, productivity, and quality of processes. The technician must be able to choose, install, and troubleshoot sensors properly.

QUESTIONS

1. Describe at least four uses of digital sensors.
2. Describe at least three analog sensors and their uses.
3. List and explain at least four types of optical sensors.
4. Explain how capacitive sensors work.
5. Explain how inductive sensors work.
6. Explain the term *hysteresis*.

7. Draw and explain the wiring of a load-powered sensor.

8. Draw and explain the wiring of a line-powered sensor.

9. What is burden current? Load current?

10. Why must load current be limited?

11. Explain the basic principle on which a thermocouple is based.

12. How are changes in ambient temperature compensated for?

13. What is the temperature range for a type J thermocouple?

14. Describe at least three electrical precautions as they relate to sensor installation.

15. Describe at least three mechanical precautions as they relate to sensor installation.

chapter

Introduction to Robotics 10

The student will gain a practical understanding of robotics. This chapter will examine the types of robots and the suitable application for each. The chapter will also consider the programming and interfacing of robots with other devices.

OBJECTIVES

Upon completion of this chapter you will be able to:
1. Define the terms *homing, axis, interpolation,* and *harmonic drives.*
2. Describe types of robot work envelopes and robot motion.
3. Describe types of robot drive systems.
4. Describe typical robot applications.

Robots are used in manufacturing for many reasons. With the costs of labor increasing and world competition forcing faster delivery, higher quality, and lower prices, robots have taken on new importance. Their use to relieve people of dangerous tasks also has increased. The environment can be dangerous because of temperature or fumes, but robots can function very well in such environments. Robots also perform repetitive operations that tend to be tedious. For example, spot welding auto bodies all day is a very tedious job. This is a boring job that can make people careless. This can reduce the quality and increase the variability of the product.

Some jobs are too difficult for humans to keep up the speed required to perform certain applications. Assembling printed circuit boards is an example. Many welding applications are inappropriate for people to do because it is difficult to reach certain areas of a part comfortably for long periods of time. Operators must wear weld masks to protect their eyes. Raising and lowering the mask every time the operator begins a weld to position the torch slows the operator and adds variability to the parts.

Robots and other automation are also appropriate when qualified labor is not available for various reasons. For example, a company's pay constraints impact the quality and availability of labor. Robots and automation provide a labor alternative. The use of automation provides an opportunity to compete in the world economy.

Automation in general impacts the type of personnel required by manufacturing. One result is a gradual drop in the need for low-skill manual labor with increased demand for manufacturing engineers and skilled technicians able to rapidly implement, use, and troubleshoot new technology.

Robots are typically used for repetitive functions such as moving and positioning parts between devices and production tasks such as welding parts together. They are fast and accurate. Robots are only a tool, however, and are a small part of a system. In general, skilled operators program them. A company can train its best welder to program the robot.

Robots are flexible tools in automated systems. They can change tasks immediately, which is becoming a more important characteristic as demands for flexibility increase.

ROBOT AXES OF MOTION

Robot axes are also referred to as *degrees of freedom*. Each axis represents a degree of freedom; for example, a five-axes robot has five degrees of freedom. Tool orientation normally requires up to three degrees of freedom, and three to four are needed to position the robot in the work space. A five-axes robot is not adequate for making complex and intricate moves such as moving the welding rod in many directions, getting into tight spots, and avoiding hitting the fixture. Six-axes robots are usually required for welding (see Figure 10-1).

Wrist Motion

A robot's wrist is used to aim the tool, or end effector, at the work piece. The wrist has three main potential motions: yaw, pitch, and roll (see Figure 10-2).

Figure 10–1 The axis of a typical six-axis robot. (*Courtesy Fanuc Robotics Inc.*)

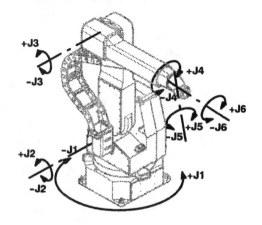

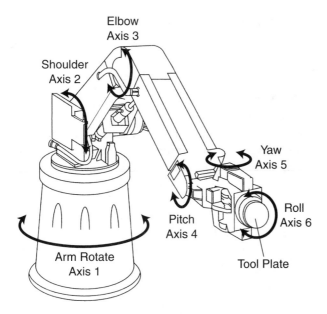

Figure 10–2 Three main motions of a robot wrist.

ROBOT GEOMETRY TYPES

The types of physical geometries of robots varies, as the following discussion indicates.

Cartesian Coordinate Robot

The Cartesian system is probably the simplest physical geometry to understand. It is usually used for assembly applications and is particularly well suited for assembly of printed circuit boards (see Figure 10-3). The main motions are on the x-, y- and z-axes. It is capable of picking components up and inserting them into

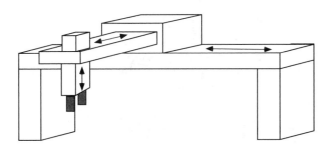

Figure 10–3 Cartesian coordinate style of robot.

boards. Its simple Cartesian coordinates make the Cartesian coordinate robot easy to program off-line.

Cylindrical Coordinate Robot

The cylindrical style robot's first axis of motion is rotate, the second is an extend axis, and the third axis is the z-axis. Other cylindrical styles are available with more axes of motion, but they all have a rotate as the first axis and an extend axis as the second axis (see Figure 10-4). Figure 10-5 shows the work envelope for a cylindrical robot. The work envelope is the area in which the robot can move.

Spherical Robot

The spherical robot has a rotate for the first and second axis of motion and an extend for the third (see Figure 10-6). This was the most popular configuration for the first hydraulic robots. See Figure 10-7 for the work envelope.

Selective Compliance Articulated Robot Actuator

The selective compliance articulated robot actuator (SCARA) typically has three vertical rotate axes and a z-axis. See Figure 10-8 for a SCARA-type industrial robot. SCARA robots are very good at working in the x-, y-, and z-orientation. They are extremely fast and particularly well suited to picking and placing electronic components.

The SCARA style is very well suited to electronic assembly because the three vertical rotational axes allow for minor location differences in the application. See Figure 10-9 for an Adept SCARA-style robot with a flex feeder. Figure 10-10 shows the work envelope for a SCARA-style robot. In Figure 10-11 a pin is to be pushed into the hole but it is misaligned slightly. It can still start into the hole because of the chamfer (angle) on the pin. A normal style of robot would have difficulty pushing the pin into the hole because the robot axis would give in many different planes. A SCARA robot would allow movement only in the x-y-plane. It would keep the z-axis in a vertical orientation. This means that the three rotational axis would allow small movement as the pin was pushed into the hole.

Figure 10–4   A cylindrical style robot whose first, second, and third axes are rotate, extend, and z.

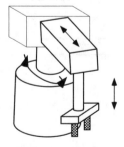

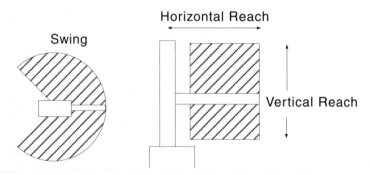

Figure 10–5 Work envelope for a cylindrical style robot.

Figure 10–6 A spherical robot.

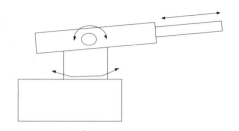

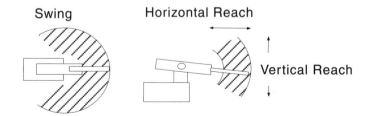

Figure 10–7 Work envelope for a spherical robot.

Figure 10–8 SCARA type robot.

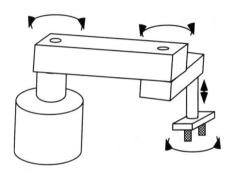

Figure 10–9 Adept SCARA-style robot with a FlexFeeder, a flexible parts presentation system that can find the parts and their orientation for the robot.

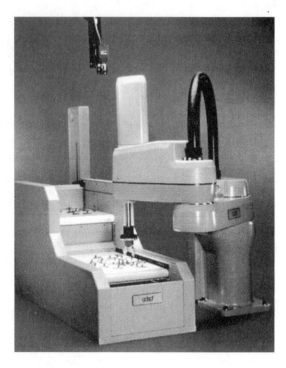

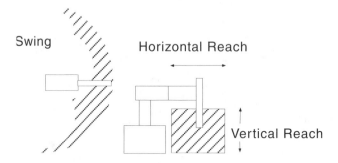

Swing

Horizontal Reach

Vertical Reach

Figure 10–10 Work envelope for a SCARA-style robot.

Figure 10–11 Simple application that involves pushing a pin into a hole.

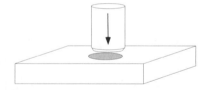

Figure 10–12 Fanuc articulate-style six-axes robot. (*Courtesy Hiromi Kugimiya.*)

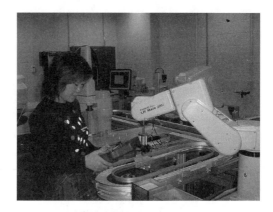

Articulate Robot

The articulate robot is the most common style of robot configuration and is most like the upper body of humans. The rotational movement of the human waist is similar to the robot's base rotate and the waist bend motion is similar to the robot's next axis. The human arm is similar to the rest of the axes. Note that humans have many more axes of motion than a robot does, which limits the robots. At least six axes are required for most applications (see Figure 10-12).

Although they are limited, articulate robots are very versatile. They can reach into tight and confined areas such as auto interiors. They can move at odd angles to weld parts together or to apply a bead of sealant in any plane. They can assemble parts and place assemblies at any angle. The articulate work envelope is called a *revolute work envelope* (see Figure 10-13).

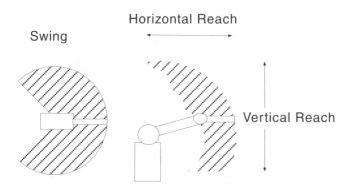

Figure 10–13 An articulated-style robot.

REFERENCE FRAMES

This section examines how the robot relates to its physical world. To do so, first imagine yourself sitting in a chair in a small room. The room represents your world coordinates. The fixed chair location establishes your right, left, front, back, up, and down. When a robot is fastened into its work cell, its *world coordinates* are established (see Figure 10-14). This world coordinate system could also be called a *world reference frame*. A robot programmer can utilize the world reference plane in a program. If we think in terms of a Cartesian coordinate system the x, y and z are fixed in relation to the robot.

User Reference Frame

In many cases, however, it would be easier for the user to establish a different coordinate frame (see Figure 10-13). Many robots allow the user to set its own reference frames, which the Fanuc Robotics Inc. calls *user frames*, to correspond with the work cell space. The user would prefer to make the robot's x, y, and z correspond to the fixtures. A user who can make the x, y, and z of the robot correspond to the fixtures can utilize CAD data to program the robot. This also makes it easier to move the robot because a move in a x, y, or z plane now follows the fixtures axis instead of the robot's world frame. This transformation to a user reference frame is very easy to accomplish on many robots. With a Fanuc robot, the

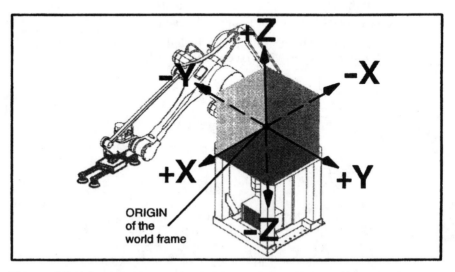

Figure 10–14   Robot's typical world coordinate system. (*Courtesy Fanuc Robotics Inc.*)

user simply teaches three points as shown in Figure 10-15. The user can create and name several user frames and then use the one that is most appropriate for the application.

User reference frames are sets of Cartesian coordinates that describe the relationship between the robot tool center point and the work space. The positional relationships between the robot and the work space are described by mathematical equations called *transformations*. The robot controller calculates the necessary joint angles that will align the robot axis to orient the tool to the work space. In other words, the robot controller aligns the robot reference frame with the work space reference plane. The transformation equations depend on the type and number of the robot's axis of motions. These equations are transparent to the programmer.

Figure 10–15 User teaches three points to a Fanuc robot to create a user reference plane. (*Courtesy Fanuc Robotics Inc.*)

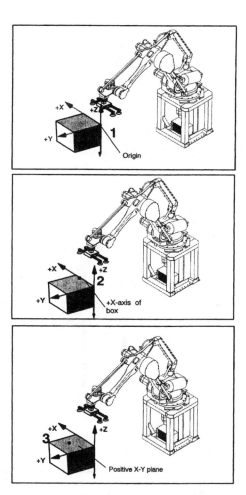

User reference frames can also be very helpful to move equipment in a cell. A new user frame can be created, and the program can be simply transformed to the new user frame and run. Fanuc also has a program shift function that teaches the robot the three points to locate the new position of the equipment, and the robot then transforms the program to the new position.

Tool Frames

Many robots can establish a tool frame. For example, in welding a circular shape, it is very important to keep the torch at the proper orientation to the weld as the robot moves the torch around the weld. A *tool frame* tells the robot what the tool's center point is. If this is not done, the robot control makes its calculations based on the robot's gripper face plate. See Figure 10-16 for an example of a tool frame.

Three methods—the three-point, six-point, and direct entry—can teach a tool frame to a Fanuc robot. The three-point and six-point methods involve bringing the tool center to a known position and recording either three or six points as the point is approached from different orientations. The robot combines these different orientations to the same point and the robot's known positions and calculates the tool orientation data to create the tool frame. Values can also be entered directly into the controller to create a tool frame.

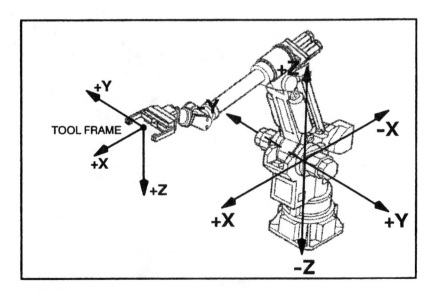

Figure 10–16 Tool frame. (*Courtesy Fanuc Robotics Inc.*)

Methods to Move the Robot During Teaching

The way the robot moves during jogging and programming can be chosen. Fanuc robots offer three modes; joint, XYZ, and tool. Each can be used to move and position the robot. The user chooses the method that is easiest to move to given positions during programming.

Joint Mode In the joint mode, each axis moves independently. A typical robot might have six axes, each of which can be moved independently. Each axis usually has two keys, a forward and backward key. The user holds one of them down and the robot moves one axis.

XYZ Mode The XYZ mode is often used to program robots. In the XYZ mode, two keys are assigned to the x-axis, two to the y-axis, and two to the z-axis. If the user chooses the +x-axis, the robot coordinates the motion of several axis to move the robot in the +x direction. The robot maintains its y and z positions while the move is made.

Tool Mode In the tool mode, the tool center point is moved in the x, y, or z direction similar to the X, Y, Z except that the tool center point frame is used for orientation.

Types of Robot Motion Just as there are three methods to move the robot when programming, there are also different move choices for the robot while actually running the programs. Fanuc Robotics Inc. offers three; joint, linear, and circular.

The joint mode is the fastest method of motion and the shortest cycle time. It is appropriate for movements around obstructions; the actual path is unimportant so long as the robot misses the obstruction. This mode is the quickest because it is not necessary for the robot to coordinate the motions of all axes together. Linear motion is used when it is important to control the path between two points as in a weld application. Circular motion is used when the robot's motion must follow an arc or circle.

We can also control how the robot moves when it reaches its programmed points. Fanuc Robotics Inc. has two choices, fine and continuous. In the *fine mode* the robot stops at each point before moving to the next. In the *continuous mode*, the robot decelerates as it approaches the next point but it does not stop at it before it accelerates to the next point. The user can enter a value between 0 and 100 to determine how closely the robot comes to the point. A value of 0 with a continuous type is a fine move. The larger the value with a continuous move, the robot stops at the farthest point from the position with the least deceleration. Continu-

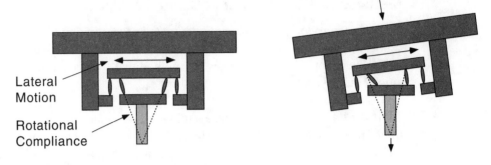

Figure 10–17 A device capable of lateral and rotational compliance.

ous mode provides smoother and quicker moves than the fine type but does not go to the exact intermediate points.

Compliance Adaptor devices also can be used in automation to allow some controlled movement in the end effector and then to return the effector to the normal position. On a robot these devices are placed between the robot and the end effector and usually allow for lateral and rotational compliance (see Figures 10-17 and 10-18).

ROBOT DRIVE SYSTEMS

Robots have several types of drive systems: pneumatic, hydraulic, and electric. Each has its own advantages and is best for certain applications. Best means that the robot can do the job and is the lowest total cost solution.

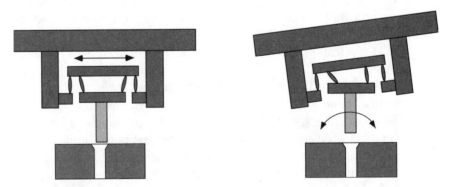

Figure 10–18 Lateral and rotational examples of compliance. Some compliance devices are capable of both.

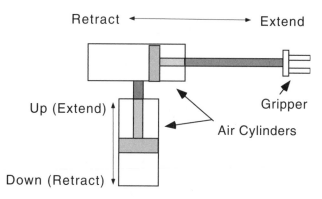

Figure 10–19 Simple pneumatic robot that can move to only four different positions with two axes, one is for up and down and one for extend and retract.

Pneumatic Robots

The simplest and least costly robot drive system is pneumatic, sometimes called *bang-bang robots* because of the way the axes operate (see Figure 10-19). It is appropriate for simple positioning tasks such as moving parts between devices. This type is inexpensive, fast, and accurate. Pneumatic robots are very limited in some ways, however, such as the number of positions to which they can move. They are also generally limited to two positions per axis, nor can they control the path taken between positions. Some pneumatic robots have flow valves to help control the speed of each axis. Some also have cushioning built into the cylinder so that each move can stop more gently.

The versatility of pneumatic robots can be improved by adding additional stops. This might permit an inexpensive, fast, simple pneumatic robot to meet the needs of a specific application. For example, an application involves basically simple moves except that it requires three positions for the z-axis (up- and down-axis): one "up" position and two "down" positions (see Figure 10-20). These positions are accomplished by adding another programmable stop and a cylinder that can extend and retract. If the cylinder is extended, it provides another hard stop for the robot. When the robot moves down, it hits this extended cylinder and stops. If the cylinder is retracted, the robot moves all the way to the bottom of its travel. A clever engineer or technician can do many things to make a pneumatic robot more versatile.

Speed and Accuracy Pneumatic robots are very fast and can be extremely accurate. An air cylinder can extend and retract very quickly and is also very accurate but can only extend and retract (see Figure 10-19). They move to hard stops and no farther. Pneumatic robots have excellent repeatability, which is the ability

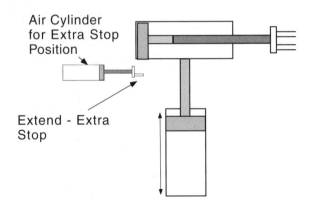

Air Cylinder
for Extra Stop
Position

Extend - Extra
Stop

Figure 10–20 An extra stop
added to make a pneumatic
robot more versatile.

of a device to return to the same position every time, because they move to a hard
stop every time. The actual stop position can usually be adjusted mechanically.

Programming Pneumatic Robots   Because of their simplicity and control
ease, pneumatic robots can be purchased with or without a controller. Control is
performed by a PLC or a device that controls and integrates the other equipment
in the cell. Pneumatic robots with controllers are normally sequential types that
can be used to pick up a piece in one position and move it to another. The sequence typically is the same every time. This type of controller is programmed by
using the robot to make each move and using a teach key to make the controller
remember the move.

Programming can be more complex if a different device such as a PLC is used
for control. Sensors and other devices provide feedback to the PLC, which can then
make a decision and turn on outputs to cause the robot to make appropriate moves.
A PLC or other complex controller is not limited to one sequence of moves.

Limit Switches   Limit switches provide robot axes with feedback to ensure that
the robot actually makes the move it is commanded to make. Proximity sensors
are usually used on pneumatic robots to do this. A *proximity sensor* is positioned
at each end of the cylinder's travel. When the cylinder extends, it closes the extend switch, which the PLC can see. When it retracts, it closes the retract switch,
which the PLC can then see. If the PLC commands the cylinder to extend, it
should easily see the extend limit switch.

Pneumatic Robot Maintenance and Troubleshooting   Since pneumatic robots are really just air cylinders, maintenance is very similar to that for any pneumatic system. Air lines can become loose or worn and leak air.

Solenoid valves, which consist of moving parts and a coil, can malfunction or
fail. Solenoid valves can be taken apart and cleaned very easily. The coils can be

easily tested by continuity and then applying power. Spare parts for solenoid valves should be kept in stock so that they can be quickly repaired and the robot can be put back into service.

If a limit switch fails, it does not provide feedback to the controller, so the sequence stops. Limit switches can be tested by looking at the input LEDs on the PLC's input module. If a robot axis is in the extend position, the extend input to the PLC should be on. The robot can then be moved to a different position to check that sensor. Such checks must be performed very carefully and not in the robot work area. If the robot is waiting for an input before it makes a move, the robot immediately makes the next move, which could be very dangerous for the operator. For this reason, robot maintenance should be performed with electric, pneumatic, and other equipment locked out.

A pneumatic robot that seems to be losing accuracy is experiencing wear or its hard stops need to be adjusted.

Hydraulic Robots

Hydraulic drive systems for robots, among the first robots developed, were capable of moving heavy loads and working at fast speeds. Some of these robots work in very hot environments such as foundries and diecasting plants to place and remove very hot parts from machines. Hydraulic robots are good for painting and moving heavy objects, and they can make very smooth moves. They are also good for dangerous applications where an electrical spark could cause an explosion.

Electric Robots

Robots with electric drive systems are very versatile, fast, and accurate. Electrical robots can use either AC or DC motors to provide their motion and typically are available with four or more axes of motion. Most robots have four or more axes. The robot controller typically contains the drives and the robot program; there is one drive for each axis of motion (see Figure 10-21). Each step of the program contains position and speed information for each axis of motion. The controller calculates speeds and distance of each move and then outputs proper voltage to each axis motor drive. The magnitude of the voltage corresponds to the desired velocity, and its polarity corresponds to the direction of travel.

The drive compares (sums) the velocity reference command (voltage) from the control device with the feedback from the tachometer. The drive then outputs an error signal, which is the difference between the command and the feedback. The drive also converts the small voltage and current to larger voltage and current to drive the motor.

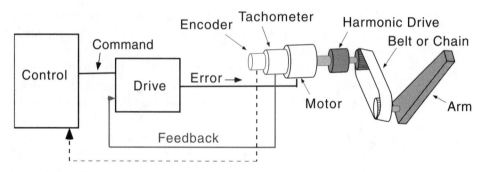

Figure 10–21   Typical axis of motion for a robot.

Meanwhile, the encoder feeds position information back to the control device, which monitors the actual and commanded positions. If a wide variation occurs during a move, the controller assumes that the arm has quit moving and inhibits the drive. This is a safety feature. The control device also controls the drive's speed of acceleration and deceleration, called *ramping* (see Figure 10-22). The controller ramps the velocity command up to the required speed, monitors the actual position of the drive, and then ramps the velocity down as the axis approaches the commanded position.

The actual drive position during a move is never exactly the same as the commanded position. The reasons for this include backlash in the drive linkage, mass

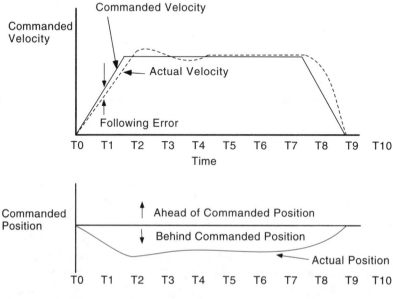

Figure 10–22   Example of ramping.

of the moving object, friction, and delays, causing the actual position to lag the commanded position. This error is a *following error*. The control device constantly monitors the amount of following error during a move and inhibits the drive if the following error goes beyond a certain value. In Figure 10-22, the top of the graphic shows the commanded velocity and the actual desired velocity. The actual velocity lags the desired velocity for some time. The bottom part of the graph shows the actual position over time. If the drive were perfectly in position, it would be on the horizontal line. If it is behind position, it is below the line. Note that because of the following error, the actual position is behind the commanded position until the end of the move when it catches up.

Homing Process

When a robot is initially powered up, it does not know what position it is in. It must be *homed* to establish its own position. Many robots, such as CNC machines, must be "homed" before operation. To do this, the operator must move the robot to a position in each axis that will be safe for homing because if a fixture, wall, object, or person is in the way, the robot will hit them when the axis begins to move toward its home position.

Each axis of a robot consists of a motor and an encoder for position feedback. Robots that have absolute encoders do not need to be homed, but most equipment has incremental style encoders and must be homed. When the robot is powered, it does not know its position but begins to move toward home, moving each axis until the robot encounters the home switch of that axis (see Figure 10-23). The purpose of the switch is to let the robot know that it is close to home. The home switch will move the robot to within one revolution of the encoder. When each robot axis finds the home switch, the robot starts slowly moving in the other direction until it finds the home pulse on the encoder. Then it knows exactly where it is. Many newer robots utilize absolute encoders and do not need to be homed when started.

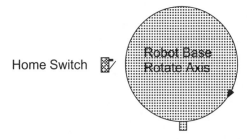

Figure 10–23 One axis of a robot. When a robot is homed, each axis moves slowly in one direction until it finds its home switch. This provides a rough position for a robot. The robot then moves each axis in the opposite direction to find the home pulse on the encoder.

Harmonic Drives

A harmonic drive is a unique and very interesting device that can create a very large gear reduction in a very small space. It consists of three main parts: a flexible can, a solid ring, and a wave generator. The flexible can has teeth on the outside called the *flexible spline* (see Figure 10-24). A solid ring with teeth on the inside is usually held in a fixed position on the outside of the can so it cannot move. The wave generator can be thought of as an ellipse that is attached to the motor shaft inside the flexible can. As it turns, the wave generator forces the flexible can (spline) to distort. The can's distortion forces a few teeth on two opposite sides of the flexible spline to mesh with teeth in the fixed ring.

The secret to the harmonic drive's gear reduction is that the fixed ring and the flexible spline have a different number of teeth. The ratio of the reduction is based on the number of teeth on the fixed ring and the difference in the number of the teeth between the fixed ring and flexible spline. For example, one may have 200 teeth and the other 201 teeth. This is a ratio of 200 to 1. If the fixed ring had 300 and the flexible spline had 298, the ratio would be 150 to 1. The wave generator turns inside the flexible spine and distorts it, causing the mesh point for the teeth to move around the fixed ring (see Figure 10-25). In this case, for every revolution of wave generator, the flexible spline moves one tooth (the difference in the number of teeth between the fixed ring and the flexible can) for a 200-to-1 reduction. The motor must move 200 revolutions to make the flexible spline move one rotation. The movement is very smooth, which allows the axis to be more stable. The device is very accurate and experiences very little wear and almost no backlash. This allows very small high-speed motors to move heavy loads.

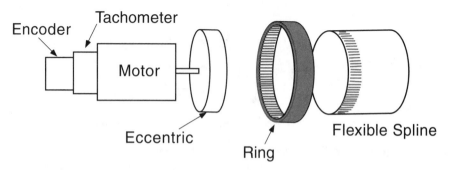

Figure 10–24   Main components of a harmonic drive.

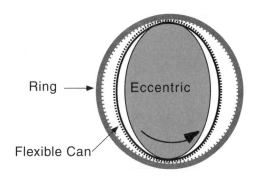

Figure 10–25 How a harmonic drive works.

MECHANIZED LINKAGES

Robots have three different styles of mechanical linkages: chain, timing belt, and gears. Chain drives utilize a sprocket on the motor shaft and a sprocket attached to the axis with a chain around both. This arrangement is very common and works well but provides a certain amount of backlash. Chains must be retightened occasionally or the robot will begin to lose repeatability. Timing belts are special belts called *synchronous* or *timing belts*. They are belts with teeth that mesh with sprockets. An idler pulley normally is used to adjust tension on the belt. Geared linkages for some robot axes are adjustable to minimize backlash for accuracy.

ROBOTIC ACCURACY AND REPEATABILITY

Accuracy refers to a robot's ability to move to a programmed position often determined by off-line programming. If a robot is programmed to move to one point in its work space, its accuracy is how close to the programmed point it can move. In shooting at a target (see Figure 10-26), accuracy is a measure of how close the shots come to the center of the target.

    Repeatability is a measure of how precisely a robot can return to a given position. Robotic repeatability is always much better than robotic accuracy. For ex-

Figure 10–26 Example of accuracy.

Figure 10–27   Example of repeatability.

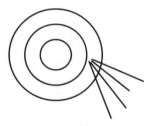

ample, the repeatability of a pneumatic robot is almost perfect because it hits hard stops at the end of each axis of travel. In the case of target shooting, repeatability measures how closely grouped the shots are (see Figure 10-27).

Types of Robot Programming

Robots are easy to program. Two types of programming, teach type and language type, are used. Some robots have a combination of the two. A robot program is very sequential. It is a series of steps, each of which contains a move and speed and other information such as required inputs or output signals or time delays.

Teach Type Programming   The teach type robot is programmed by moving the robot to the desired position. The programmer then inputs the information for that particular step and the desired speed. If an input is required before the next move is made, it is entered in the step. The programmer also enters any outputs required such as the gripper or welding output or other outputs that send a signal to other equipment.

A derivation of the teach type robot is the *lead-through type*, which is particularly suited to painting applications. The best painter in the shop would put the robot into teach mode and then grab the end of the arm and paint a part. The robot remembers the speeds and motions that have been taught and can repeat the motions and speeds.

Language Programming   The other type of robot programming uses a language. It might be very similar to a Pascal or C language program with some special functions for moves, loops, and palletizing routines. A programming language makes the robot more versatile but is somewhat more difficult to learn to program. The use of a programming language allows the program to be written off-line on a computer and downloaded to the robot. Typical robot languages have if/then/else statements for loops, while and do loops, do until loops, and case statements.

Following is a partial listing of a program for a Fanuc robot. The commands are quite easy to understand and use. Note that each line has been commented to explain its function.

! Comment—Line 1 turns digital output 7 on.

1. **DO{7} = ON**
   ! Comment—Line 2 is a move command. The command is to move in joint mode to position 1 at 100% of the programmed speed.
2. **J P[1] 100% Fine**
   ! Comment—Wait for digital input 1 or 3 or 4 before continuing the program.
3. **WAIT DI[1] = ON OR DI[3] = ON OR DI[4] = ON**
   ! Comment—Turn robot output 1 on.
4. **RO[1] = ON**
   ! Comment—The command is to move in joint mode to position 1 at 100% of the programmed speed.
5. **J P[1] 100% Fine**
   ! Comment—The next line is a conditional statement. If digital input 5 is on, the program will jump to the section of the program called LBL 2. It will skip all of the program lines up to LBL2.
6. **IF DI[5] = ON JMP LBL[2]**
   ! Comment—The command is to move in joint mode to position 3 at 100% of the programmed speed.
7. **J P[3] 100% Fine**
   ! Comment—The command is to move in joint mode to position 4 at 100% of the programmed speed.
8. **JP[4] 100% Fine**
   ! Comment—this is label 2. Labels are a way to identify sections of a program and can be used to identify where a program should jump to
9. **LBL[2]**
   ! Comment—turn robot output 1 on.
10. **RO[1] = ON**
    ! Comment—wait until digital input 2 is on.
11. **WAIT DI[2] = ON**
    ! Comment—The command is to move in joint mode to position 9 at 100% of the programmed speed.
12. **JP[9] 100% Fine Offset**
    ! Comment—Wait 2 seconds before continuing the program.
13. **WAIT 2.00 (sec)**
    ! Comment—The command is to move in joint mode to position 10 at 100% of the programmed speed.
14. **JP[10] 100% Fine Offset**
    The programmer would then teach each of the points in the program by moving the robot to the desired locations and recording each. Note that in this

> program points were given numbers as names. They could have been named alphabetically instead.

Many of the languages also provide special functions for operations such as palletizing. The operator simply supplies information about the number of rows and columns and location of corners. Welding robots have special functions available to make weaving easy, and others have functions that help integrate a vision system. The vision system can help find the parts and or inspect the parts and then share this information with the robot.

Communications   There are many types of communications available for robots today. The most limited is RS-232, which provides basic communications capability. RS-232 has historically been used to connect a robot to a microcomputer for programming and uploading/downloading programs. Its speed and noise immunity are quite limited which restricts the distance over which communication can take place.

Robots can now be networked with Ethernet, DeviceNet, ControlNet, and several other systems. These communication networks are fast and powerful and are readily available for computers, PLCs, and other devices. They make integration an easy task.

ROBOTIC APPLICATIONS

There are disaster stories about companies that purchased robots that failed miserably in their application. Many of these robots ended up sitting in warehouses gathering dust, often because the application was not appropriate for the robot or too complex. One of the first things companies learn is that robots are very unforgiving. They also quickly learn that they need to improve the quality and consistency of their part production.

They realize that parts have to be positioned very accurately and consistently for a robot to be able to pick them up consistently.

Welding

Arc Welding   Arc welding is very appropriate for a robot. The robot uses an electric arc to generate the heat to fuse metals together with wire as a filler material. Robotic arc welds are much more consistent than human welds. Robots also use a flame-cutting torch or laser as a tool to cut out shapes in metal. For example, motorcycle manufacturers use a laser tool to cut shapes into exhaust heat shields.

Manufacturers make parts that vary very little but must provide good fixtures to hold the parts for the weld. The fixtures must accurately locate the parts and position them for the weld. Parts must generally be held accurate to one-half of the material thickness to be a candidate for robotic arc welding.

Some unique robotic methods are being used to compensate for part inaccuracies. Vision has been tried but has not been very successful because of the dirty environment that welding creates. Another method has been to sense the current during the weld. A weld generally uses a weave pattern to weld pieces together. By sensing the current, the robot path can be adjusted automatically to compensate for inaccuracies and gaps.

Spot Welding   Spot welding fuses two pieces of metal together, generally along a seam. It is an excellent application for robots. An electrical current passing through the parts creates the heat. A spot welder is a tool that resembles the human thumb and a finger. The parts are positioned in a fixture, and the robot closes the spot welder (thumb and finger) over the desired area and clamps it. Current is passed through the metal, creating heat that fuses the two pieces together. The spot welding attachment is very tolerant of positional inaccuracies, and the robot can do the welding consistently and accurately.

Investment Casting

Investment casting is an application that primarily involves part handling. The robot picks up a wax impression of the parts to be made such as steel golf club heads. The parts are often in a treelike configuration. The robot then moves the wax tree part into a slurry of fine refractory material, which dries on the wax parts. More coats of the refractory material are added. The robot spins the molds to ensure consistent thickness of refractory material. The robot is finished at this point, and the lost wax process is used to mold the parts.

Forging

Forging is the process of hammering metal into a required shape, generally by presses. Forging is done at high temperatures so that the metal is in a malleable state. This heavy, dangerous, and noisy work is well suited to a robot. The robot moves the part from furnace to press and then to an outgoing bin.

Die Casting

Robots act as part movers in a die cast application, which typically is performed in a hot, dangerous environment. The robot removes very hot finished parts from a die cast machine. The robot removes the part and puts it in a press that trims the part. This removes excess material and flash from the part.

Painting/Sealing/Gluing

Spray painting is a good application for robots. As noted earlier, hydraulic robots are generally used for spray painting applications because they do not generate sparks that could ignite the fumes in a paint booth. Hydraulic robots also apply paint very smoothly. Robots also apply glue and/or sealants to surfaces such as an automobile windshield. One robot applies sealant to the rim of the window and another robot puts the windshield into the car.

Loading/Unloading

Diecast, forging, and loading/unloading applications are load/unload applications. CNC machines can be fully automated with the addition of a robot that loads parts into the machine and unloads parts from it.

Assembly

Robots often perform assembly operations. Robotic machinery often assembles printed circuit boards. They also assemble other parts, such as motors and disk drives.

Finishing/Deburring

Robots' ability to make complex three-axis moves is very useful for finishing and deburring applications. A robot is a natural choice to move a complex part such as a rotor with many sharp edges and deburr the part. Robots hold parts against belts to shape them or remove burrs.

Inspection

Robots are adept at inspection tasks. One good example is in automobile inspection. They can move cameras into position to move and triangulate sizes. They can do this very quickly right on the production line. Robotic inspection enables every car to be inspected instead of a random inspection. It also eliminates the need for an inspection fixture to hold and locate the part. The vision system finds the car and adjusts to its position for the inspection. This avoids disrupting the assembly line for hours of inspection.

Automated Guided Vehicles (AGVs)

Automated Guided Vehicles (AGVs) automatically move parts and materials through a plant. They are controlled in two ways. One way is for the robot to sense a wire in the floor. The other method is by radio frequency, which can transmit commands to the AGV. These methods are quite often used in automated storage facilities. They are guided and controlled by a central computer.

Automated Storage/ Automated Retrieval (ASAR)

An automated storage/automated retrieval (ASAR) system is really a rectangular style robot. It is used to load or unload materials into bays in storage facilities. It can be extremely large. A computer optimizes the storage and retrieval to ensure the security of inventory and the use of the oldest first. It also helps optimize the use of storage space.

APPLICATION CONSIDERATIONS

The application usually dictates the type of robot to be utilized. The best choice is usually the most inexpensive one that will accomplish the task. Some things to be considered are type of application, speed required, complexity of task, type of motion required, need for flexibility, weight involved, and hazards. Some applications may not require a robot. A linear drive (axis) or two may be able to automate a simple part of an application very cost effectively. The main types of motion are SCARA or articulate.

Electronic assembly often dictates the use of a selective compliance articulated robot actuator (SCARA) style robot because it is very good working in x-, y-, and z-axes. These robots are very fast and very good at picking and placing components in an x, y, z orientation.

Painting, applying a sealant, deburring or polishing, machine loading/unloading, or welding applications usually dictate an articulate style robot. This robot is capable of working in any orientation. Electric robots are the most common choice, but hydraulic robots are also good for painting applications because they are very smooth.

Pneumatic robots are a great choice for simple repetitive tasks. They are generally more suited to applications that require little flexibility.

Hazardous and very heavy or hot applications may require a hydraulic robot. If the robot is working in an area with flammable materials, the hydraulic type is a good choice. The robot control can be located outside the area and only the actual robot is in the area. There is no danger of sparks igniting the flammable materials. It should be noted, however, that electric robots have also been greatly improved and spark proofed for hazardous environments.

ROBOT INTEGRATION

Some robots are programmed using a specialized robot language. These robots, which have digital (on/off) inputs and outputs (I/O) available, are very easy to connect to control devices such as PLCs, computers, or other devices. Most robots

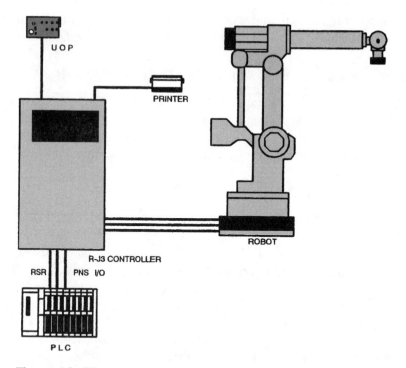

Figure 10–28   A robot communicates with a PLC. A digital output from the robot becomes an input to the PLC. A digital output from the PLC becomes an input to the robot. (*Courtesy Fanuc Robotics Inc.*)

wait for an input to tell it what task to do and then performs the task and sends an output to a cell controller to tell the control device that it has completed the task and is waiting for another one (see Figure 10-28).

END-OF-ARM TOOLING

Much design goes into the tooling on the end of a robot. Grippers activated by a pneumatic cylinder are used to pick up and place one size of a part. Often, however, many parts of different shapes and sizes must be picked up and moved. Different areas of the gripper can be configured to pick up different parts. Grippers can also be designed to work in a vacuum. Many types of suction cups are available. Figure 10-29 shows an example.

Weld torches are also a type of end-of-arm tooling. Figure 10-30 shows an example of an articulate robot with a weld torch installed for wire welding. Other

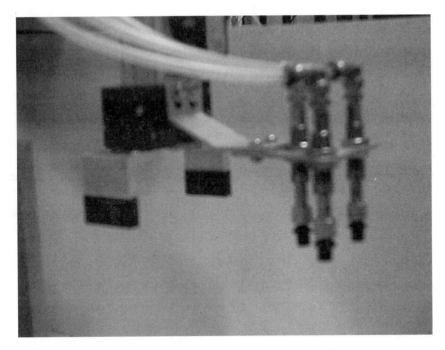

Figure 10–29 This gripper has three suction cups. The arm also has a simple open/close gripper.

types of tooling include automatic screw guns widely used in assembly operations to dispense a fastener and then tighten it.

In complex applications, robots may use several tools. The robots in these applications change their tooling as needed to accomplish their tasks. Air lines and electrical connections are standardized and reconnect to each new tool as it is loaded.

ROBOT MAINTENANCE

Robot maintenance should first be preventive maintenance. Preventive maintenance should be planned to avoid interfering with production. The manual for the robot or other automated equipment discusses procedures for preventive maintenance. Proper preventive maintenance can reduce the number of breakdowns. One of the most important things is replace the backup battery on the recommended schedule. A loss of battery power causes the robot to lose valuable information. Robots with absolute encoders will lose positional information and will need to be retaught their home position.

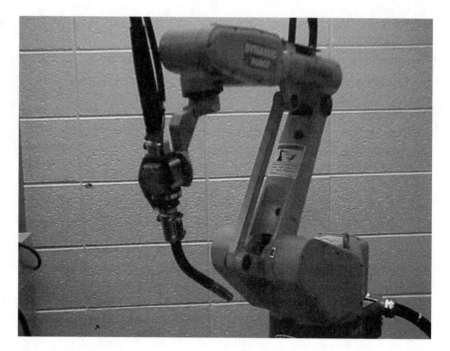

Figure 10–30   Weld torch for wire-feed welding application.

Robot Troubleshooting

Newer computerized equipment, such as robots, and CNC equipment have troubleshooting functions built into the control. The technician can have the control unit run diagnostics to help find a problem. The manual for the equipment normally has a troubleshooting sequence to lead the technician to the problem. In some cases the technician must call the equipment manufacturer for advice. The manufacturer's expert tells the technician what to check. This usually identifies the problem very quickly and usually at no cost.

Crashes

Crashes can cause physical and electrical problems. The robot control function can usually help identify the problem. Modern controls provide error messages with specific suggestions for correcting the errors. Serious robot crashes can cause the axes to be misaligned. The operator must test the program of a robot restarted after a major crash very carefully the first time. It may be best to single step the program at very low speed. If the axes have become misaligned or have shifted, slow movement will prevent more damage.

A severe crash can cause physical damage to the robot. Chains, belts, keys, or other linkages can be broken. In some cases a part of the robot housing may even break. The electrical protection that is built in can sometimes prevent serious damage. Current limits built into the drive of each axis should cause the drive to inhibit motion if the current limit is exceeded. When the drive is reenabled, the robot should run again. In the case of a rapid, severe crash, the current limit may not protect the robot from further damage.

One of the more common electrical problems after a crash occurs with the drive output transistors. A large, quick overcurrent condition caused by a crash can ruin one or more of the output transistors in the drive (each axis of motion has one drive). Many drives have some built-in troubleshooting device such as an LED on the front of the drive that should be to indicate that the drive was not harmed.

When a robot or control device is down, production normally stops. Downtime is incredibly expensive. To avoid long downtime, spare parts for important equipment and parts that are most prone to fail must be kept in stock. A maintenance kit that contains the parts most likely to fail are generally available when the robot is purchased. A spare motor, tachometer, encoder, electrical boards, and a spare drive should usually be kept. A robot may have four or more axes, all using the same drive.

Lost time in industry is too expensive for a technician to take time to repair a board or drive when the equipment is down. The part should be removed and replaced with a new one. The operator can sometimes try to repair the item later, but it is usually best to send the old one to the manufacturer who will send a new one and repair the old one, usually at a very reasonable cost.

Robots and other computerized equipment have multiple axes that often contain the same devices. Sometimes it is possible to swap components to isolate the problem. For example, the drives may be the same. If one drive has the axis fault LED on, that drive might be exchanged with another to see whether the light stays on in the new position (axis) or if it comes on for the drive that was moved to the problem axis. A light that stays illuminated on the same drive indicates a drive problem. If the axis fault light on the good drive that was moved to the problem axis comes on, the drive is not the problem. The technician should check the manual first to ensure that the components are compatible before they are swapped.

QUESTIONS

1. List at least four reasons that robots are used in industry.
2. Describe the terms yaw, pitch, and roll.
3. What is a reference frame?

4. Describe a Cartesian style robot.

5. What is a SCARA style robot?

6. What is a compliance device used for?

7. Name three types of robot drive systems.

8. List at least two advantages of pneumatic robots.

9. List at least two advantages of electric robots.

10. If a robot has incremental encoders on each axis, they do not need to be homed. True or False?

11. What is the purpose of a harmonic drive?

12. What is the difference between accuracy and repeatability?

chapter 11

Introduction to Fluid Power Actuation

Pneumatics and hydraulics are very prevalent in industry. The basic understanding of such systems is crucial in design, troubleshooting and maintenance of automated systems.

OBJECTIVES

Upon completion of this chapter you will be able to:
1. Understand the basic principles of fluid power.
2. Define the terms solenoid, spring-return, double-acting, and flow valves.
3. Design simple fluid power applications, including calculating sizes, and choose appropriate equipment.
4. Explain the types and operation of simple valves.

FLUID POWER ACTUATION

Fluids such as air and oil have long been used to do work. Pascal's law, which is the basis of fluid power, states that pressure applied on a confined fluid is transmitted in all directions and acts with equal force on equal areas and at right angles to them (see Figure 11-1). Consider a container similar to the one in Figure 11-1. If the cork is pushed far enough into the bottle, the bottle will break. This happens because the liquid is incompressible and transmits the force from the stopper through the whole container. This causes a much higher force on a much larger area than the cork. Thus, the glass container can break with relatively small pressure.

A simple example of this law applied to a practical application involves a simple hydraulic press (see Figure 11-2). The left part of the press has a small input area and force is applied to this side. This causes the same pressure to be applied throughout the press. The forces are proportional to the area of the pistons. Thus, a

229

Figure 11–1 Illustration of Pascal's law. The pressure is exerted in all directions and with equal force at 90 degrees to the surfaces.

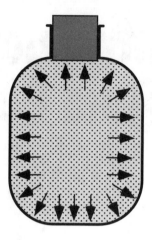

5 pound force can lift a 50 pound load if the output piston has 10 times the area of the input piston. Figure 11-3 shows the comparison between volume and distance.

Pressure

Pressure is expressed as atmospheric, gauge, and absolute. *Atmospheric pressure* is the force exerted by the weight of the atmosphere on the earth's surface. The weight of the atmosphere at sea level is 14.7 pounds per square inch. Atmospheric pressure can also be expressed in inches of mercury. A mercury barometer

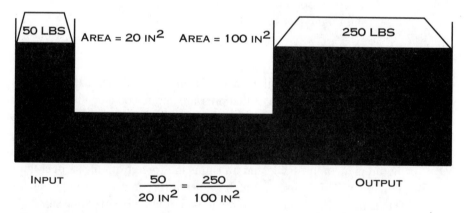

INPUT $\dfrac{50}{20 \text{ IN}^2} = \dfrac{250}{100 \text{ IN}^2}$ OUTPUT

Figure 11–2 Some basics of hydraulic leverage. The area of the input side of the press is 20 square inches, and the area of the output side is 5 times larger (100 square inches). This means that we can lift 5 times as much weight as the amount of pressure we apply at the input.

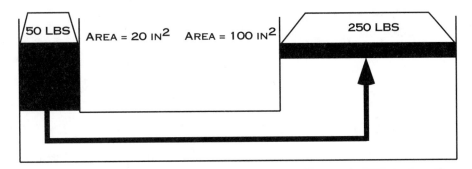

Figure 11–3 If we move a given volume of fluid on the input side, the same volume is moved on the output side, resulting in a smaller move. In this case while we could lift 5 times our input pressure, we lose distance. The output in this case only moves one-fifth a the distance.

calibrated in inches of mercury can be used to measure atmospheric pressure. The 14.7 pounds at sea level represent one atmosphere; it is equal to 29.92 inches of mercury. *Gauge pressure* is the pressure above atmospheric pressure. Most pressures are gauge pressure. *Absolute pressure* is the pressure above a perfect vacuum. It is the sum of gauge pressure and atmospheric pressure. Absolute pressure is expressed in pounds per square inch absolute (psia).

Work and Power

$$\text{Work} = \text{Force} \cdot \text{Distance}$$

Work is expressed in foot pounds. If 5 pounds are lifted 10 feet, the work actually done is 50 foot pounds (5 pounds · 10 feet). Imagine walking up a flight of stairs. The work done equals your weight multiplied by the height of the stairs. If you run up the stairs, you are doing the same amount of work but at a faster rate, called *power*. It takes much more power to run up a flight of steps than to walk up them even though the same amount of work is done.

$$\text{Power} = \frac{\text{Force} \cdot \text{Distance}}{\text{Time}}$$

Power is usually measured in horsepower or watts. One horsepower equals 33,000 pounds lifted in one minute. One watt equals 1 newton lifted one meter in one second. One horsepower is equal to 746 watts. Pressure equals the force of the load divided by the piston area (see Figure 11-4). The flow in fluid power systems is measured by the flow rate of gallons per minute (GPM). Speed and distance are indicated by the flow. Force is indicated by pressure.

Figure 11–4 Formulas for pressure.

$$P = \frac{F}{A} \qquad 1W = \frac{1N\ m}{sec.}$$

$$A = \frac{F}{P}$$

$$F = P * A$$

SIMPLE APPLICATION

Let's try a simple application. We need to lift a 500-pound load 10 inches. First we must select an actuator. In this case, a simple linear actuator would do the job. The load must be moved 10 inches, so we must choose a cylinder that has at least a 10-inch stroke. The area of the cylinder depends on the weight of the load and the pressure that will be used in the system. For this example, assume that our maximum operating pressure is 100 pounds. A 5-square inch cylinder would lift 500 pounds. Our pressure is 100 multiplied by the 5-square inch cylinder would be 5000 pounds. This would not be a good idea, however. We should provide a margin of error. An 8-square inch cylinder would lift the load at 62.5 pounds of pressure and provide a margin of error. See Figure 11-5.

TYPES OF ACTUATORS

Linear Actuators

Linear actuators (cylinders) are classified according to their construction and method of operation. The single-acting and double-acting cylinders are the basis for all other cylinders. A *single-acting cylinder* has a spring to return the piston rod (see Figure 11-6a). Air or oil is used to extend the cylinder and when the pressure is released, the spring returns the piston. A *double-acting cylinder* uses air to extend and retract the piston (see Figure 11-6b).

Cushioning can be added to decelerate the piston at one or both ends of travel. The amount of cushioning is adjustable with a needle valve that controls the flow rate of air or oil as it escapes the cylinder. The cushioning is designed to work at

Figure 11–5 Simple application of lifting the load 10 inches.

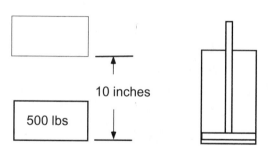

10 inches

500 lbs

Figure 11–6 Various cylinder symbols.

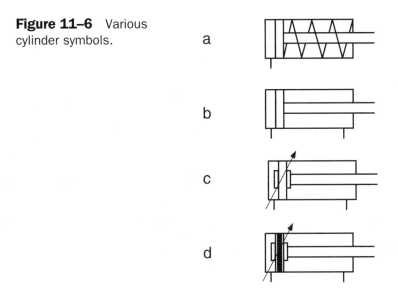

the end of the travel. The cylinder can move full speed between the endpoints of travel. This is especially useful for moving heavy loads. The cushioning allows controlled deceleration before a stop. See Figure 11-6c for a double-acting cylinder with adjustable cushioning at both ends.

Both single-ended and double-ended cylinders are available. A *single-ended cylinder* produces unequal forces on the extend and retract stroke. On the extend stroke, the fluid pushes against the whole surface of the piston. On the retract stroke, the area is reduced by the diameter of the piston rod. A double-ended cylinder has equal pressure on the extend and retract stroke. Cylinders are also available with piston rods that will not rotate.

Rodless cylinders are also available (see Figure 11-7). These cylinders have a housing on the outside attached to the piston. The outside of the cylinder has a

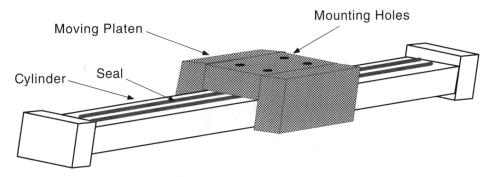

Figure 11–7 Rodless cylinder.

Figure 11–8 Principle of a rodless cylinder.

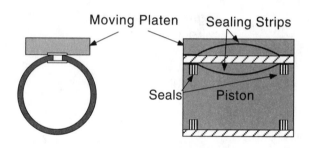

seal so that air fluid does not leak. Rodless cylinders are particularly useful where space is tight. Instead of having a rod attached to the piston that extends out of the cylinder, they have a platen that is attached through one side of the cylinder (see Figure 11-8). The cylinder moves along the outside of the cylinder. The mounting holes on the platen attach to the object that needs to move.

Rotary Actuators

A rotary actuator is a motor with a rotor that can be pushed by air to a stop in either direction. A single-rotation type can only make a revolution of 360 degrees or less. Imagine that we need to design a rotary table that stops in two positions. A rotary actuator would be a very economical way to do this. A rotary actuator uses two air cylinders (see Figure 11-9), each of which moves a rack. A pinion is attached to the rotor. The rotor is moved by the action of the cylinders pushing the rack and moving the pinion. The stops on a rotary actuator are adjustable.

Air-over-oil systems are available if smooth controlled movement is required. In these systems oil actually moves the pistons, but air pressurizes the oil. This allows the speed and smoothness of the move to be controlled because oil is non-compressible. Rotary actuators are also available for pneumatics. The two types are: motor and single rotation. The motor type is an air-operated motor.

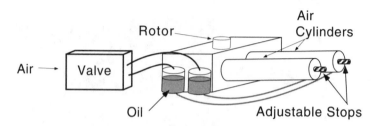

Figure 11–9 Air-over-oil system with a rotary actuator. Rotary actuators are available for pneumatic or oil systems.

Feedback on Cylinder Position

It is sometimes necessary to have feedback as to the actual position of the piston. Magnetic sensing can be added to give feedback on whether the cylinder is in the extended or retracted position. See Figure 11-6d for the symbol for a double-acting cylinder with adjustable cushioning and magnetic sensing.

Actuator Speed

Actuator speed depends on the size of the actuator and the flow into it. To calculate the speed, the volume to be filled in order to cause the needed amount of travel must be considered. The speed of an actuator can be controlled with an adjustable flow valve, which can restrict the flow rate of air. Restrictors are installed at the outlet of the cylinder to slow the flow of air out of the cylinder. The relationship is shown in Figure 11-10.

Air Pressure

Figure 11-11 shows a variety of symbols used for air supply and control devices. Pressure is usually shown with the first symbol. A variety of filter/regulator units are available to regulate the air pressure as well as clean and lubricate the air. Some units remove oil from the air for certain applications.

Directional Valves

Directional valves are used to stop, start, and control the direction of fluid flow. They can be classified according to the following characteristics:

Figure 11–10 Formulas used to calculate speed.

$$volume\ /\ time = speed * area$$

$$speed = \frac{volume\ /\ time}{area}$$

$$area = \frac{volume\ /\ time}{speed}$$

$$v/t = in^3/minute$$
$$a = in^2$$
$$s = in/minute$$

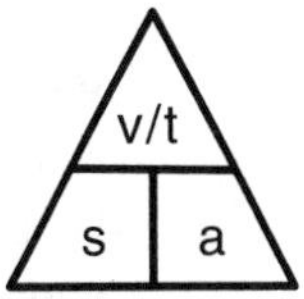

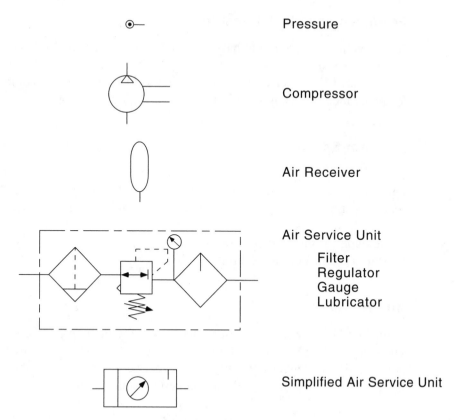

Pressure

Compressor

Air Receiver

Air Service Unit
  Filter
  Regulator
  Gauge
  Lubricator

Simplified Air Service Unit

Figure 11–11 Air supply and control device symbols.

The method of actuation: Manually, mechanically, or by pneumatic, hydraulic, electrical power, and a combination of actuation such as pneumatic and manual.

The number of flow paths they provide: two-way, three-way, and four-way valves.

The type of connection: pipe thread, straight thread, flanged and subplate, or manifold type mounting. Valves can also be classified by the internal type of valve that is used: poppet (piston or ball), sliding spool, and rotary spool.

Valve Symbols Valve positions are represented by squares (see Figure 11-12). The number of squares represents the number of switching positions. The one in Figure 11-12b has two switching positions. Lines in the boxes indicate flow paths, and arrows represent the direction of the flow (see Figure 11-12c). Valve

Figure 11–12 Valve symbols. a. Each rectangle represents a valve position. b. This drawing represents a two-position valve. c. A line in a box with an arrow represents flow and its direction. d. A line at a right angle represents a shut-off position. e. Inlet and outlet ports are represented by lines on the outside of the box. These lines are drawn in the initial valve position.

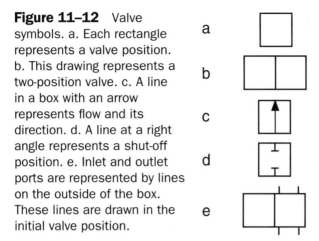

shut-off positions are represented by lines at right angles (see Figure 11-12d). Inlet and outlet ports are shown by lines (see Figure 11-12e).

Directional control valves are represented by the number of ports and the number of control positions that they have (see Figure 11-13). Normally open and normally closed are the opposite in fluid power. In electrical terms, a normally

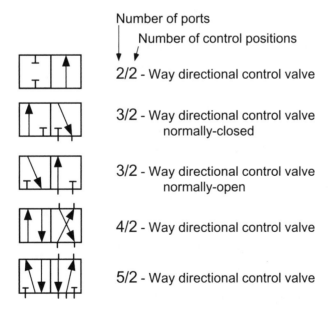

Number of ports
Number of control positions

2/2 - Way directional control valve

3/2 - Way directional control valve
normally-closed

3/2 - Way directional control valve
normally-open

4/2 - Way directional control valve

5/2 - Way directional control valve

Figure 11–13 The most common directional valve configurations.

Figure 11–14   Standard numbering and lettering system for ports.

Pressure Port - 1 or P
Exhaust Port - 3 or R
Exhaust Ports - 5, 3 or R, S
Signal Outputs - 2,4 or B, A

open valve does not pass current. In fluid power, a normally open valve does pass fluid.

Port Numbering   A numbering system identifies the ports on directional control valves. In the past, a lettering system was used. Both are shown in Figure 11-14. The pressure port is 1 or P. Exhaust ports are 3 and 5 (R and S). Signal outputs are 2 and 4 (B and A). Directional control valves are represented by the number of ports and the number of control positions that they have.

The simplest type of directional valve is a check valve, which allows flow in one direction but blocks flow in the other direction. It is made of a ball, a seat and a spring (see Figure 11-15). A light spring pushes the ball against the seat to stop flow in one direction. If the pressure on the other side is higher than the light spring pressure, the ball moves away from the seat and allows flow in that direction.

Pilot-Operated Check Valve   A pilot-operated check valve is controlled by a pilot signal that is electrical, mechanical, or fluid operated. The pilot-operated check valve operates like a simple check valve until pressure is applied through the pilot port. When pilot pressure is applied, reverse flow is permitted through the valve.

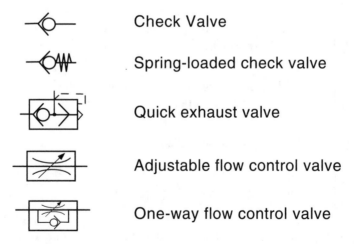

Check Valve

Spring-loaded check valve

Quick exhaust valve

Adjustable flow control valve

One-way flow control valve

Figure 11–15   Some of the more common types of check valves.

Figure 11–16 Two-way valve showing spool.

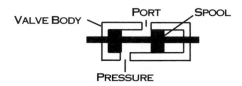

Two-Way, Three-Way, and Four-Way Valves Imagine that we need to make a cylinder extend and retract. We need a device to control and change the direction of fluid flow. These devices are called *valves*. They can be controlled manually or by pneumatics or electricity. A valve consists of a body that has internal flow passages and ports to connect to. Valves are all designed to direct flow from the inlet port (pressure port) to either of two output ports. The number of ports to and from which fluid flows determines whether the valve is a two-way, three-way, or four-way valve.

Two-Way Valves A two-way directional control valve has two ports connected to each other with passages. These passages can be opened or blocked by a spool (see Figure 11-16). The figure shows the body of the valve, the spool, and the ports. Figure 11-17 shows the two possible valve positions. In the first position, the flow path is open between the pressure port and the actuator output port. The graphic to the right of the valve is the graphic representation of this valve position. The arrow shows that the ports are connected and the direction of the flow. Note that this would be a 2/2 valve because there are two ports and the valve has two possible positions.

The second position shows the valve closed. No flow can move from the pressure port to the actuator output port. The graphic to the right of the valve is the representation of this valve position with the ports blocked by the spool. A two-way valve provides an on/off function only.

Three-Way Valves A three-way valve has three ports that can be connected by passages in the body of the valve. Figure 11-18 shows the two possible valve spool positions. Note that port P is the supply pressure, port A is connected to the cylinder, and port EX is the exhaust port. The graphical representation on the left shows that the supply port and the actuator port are connected and the exhaust port is blocked. The graphic on the right shows that the supply port is now

Figure 11–17 Two possible valve positions.

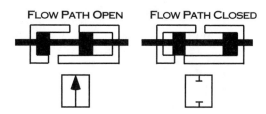

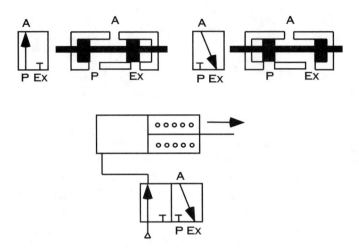

Figure 11–18 Two possible spool positions.

blocked and the cylinder port is connected to the exhaust. This valve can be used to apply or exhaust pressure to one port. This valve is a 3/2 valve because it has three ports and two positions.

Four-Way Directional Valves The four-way valve is probably the most commonly used valve in pneumatic systems. It has a pressure port (supply pressure), two actuator ports, and one or more exhaust ports. Figure 11-19 shows a four-way valve. In this figure the spool has connected the pressure port to the A port, and the B port is connected to the exhaust port.

Figure 11-20 shows the two possible positions of the spool and the graphical representation of each valve position. The left view shows the spool in the left-most position. In this position port A is connected to the exhaust port. The pressure port is connected to Port B. The symbolic representation shows the same thing. The arrows represent the connections and the direction. This is a 4/3 valve because it has four ports and two positions.

The right graphic in Figure 11-20 shows the spool in the right-most position in which the pressure port is connected to the A port, and the B port is connected to

Figure 11–19 Four-way valve.

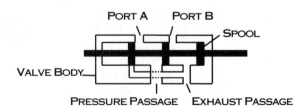

Figure 11–20 Two possible spool positions and a graphical representation of each valve.

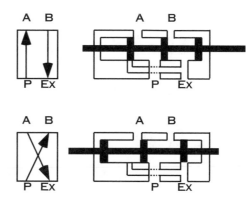

the exhaust. The symbolic representation is clearer. The arrows show that the A port is connected to the exhaust and the pressure port is connected to the B port.

We could construct a simple but very useful circuit with this valve (see Figure 11-21). This figure shows a symbolic representation of a four-way valve connected to a double-acting air cylinder. The spool is in the left position. The pressure port is connected to the A port, and the B port is connected to the exhaust. As pressure is applied to the rear of the cylinder, the pressure in front of the position can go through the B port to the exhaust. If we move the spool to the right position, the pressure is applied to the B port (front of the cylinder), and the back of the cylinder could exhaust the pressure through the connection of the A port to the exhaust.

Three-Position Valves The valves discussed to this point were all two-position valves. Three-position valves are also available in four and five port models. Figure 11-22 shows three examples of three-position five-port valves. On the left is a blocked center valve. The center position in this valve blocks all of the working ports. The left position extends the piston cylinder, and the right position retracts the piston cylinder. The center in Figure 11-22 is an exhaust center valve. When the

Figure 11–21 Simple circuit that utilizes a four-way valve.

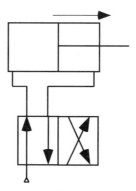

Figure 11–22   Three examples of three-position valves.

valve is in the center position, both ports are opened to the exhaust. In this case, the piston in a cylinder could float in either direction. In the left position in this type of valve, the piston in the cylinder extends. In the right position, the piston cylinder retracts. The right in Figure 11-22 is a pressure center valve. When the valve is in the center position, both ports are opened to pressure. In the left position, the piston in the cylinder extends. In the right position, the piston cylinder retracts.

Spool Position Valves   *Spool position* classification refers to the normal or deenergized valve condition. There are several types. We consider two of the types available.

One type is spring centered. In a spring-centered valve, the spool is returned to the center position by spring force. When the actuation pressure is gone, the spring returns the spool to the centered position. The spring return type is valuable in some applications if power is lost, because the valve can return the cylinder to a safe position.

Spring-return valves can be purchased as either normally open or normally closed valves, which is determined by the condition when the valve is not energized. Figure 11-23 shows a symbolic representation of four spring-return valves.

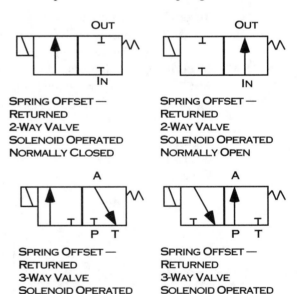

Figure 11–23   Symbolic representation of four spring return valves.

The valves in Figure 11-23 are solenoid operated. A solenoid is a coil used to generate a magnetic field to move the spool in one direction. When the coil is deenergized, the spring returns the spool. The symbolic representation of spring-return valves always shows the valve in the deenergized condition.

A second type is a two-position spring offset type valve. Its spool is normally offset to one end by a spring. It has one actuator that shifts the spool to the other location. When the actuation force is released, the spool returns to the end position.

VALVE ACTUATION

There are several ways in which valves are actuated. One way is to manually activate the valve. It is activated by the mechanical movement of a lever or a plunger. A mechanical example is a handle (lever) that an operator must move to make something happen. The plunger style is used for automatic operation. Some part of the machine pushes the plunger, which changes the valve spool position. Figure 11-24 shows examples of several types of activation.

Mechanical Actuation

Very similar to manual actuation, mechanical actuation does not involve a person. This type of valve has a plunger that is activated when it is moved by a cylinder or machine element; this movement moves the spool.

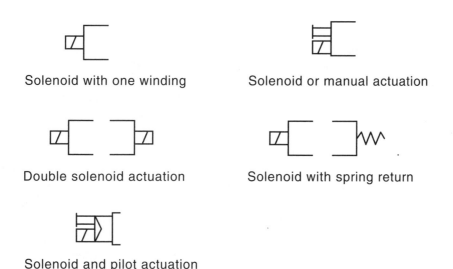

Solenoid with one winding

Solenoid or manual actuation

Double solenoid actuation

Solenoid with spring return

Solenoid and pilot actuation
with manual override

Figure 11–24 Diagram of several common valve actuation types. A valve can have more than one method of actuation.

Fluid Power Actuation

Fluid shifts the valve spool in these valves. A pilot is another port on the valve. If pressure is present on the pilot valve, the piston shifts and allows flow. When the piston shifts, it moves the spool, which changes the flow direction.

Electrical Actuation

Valves that provide electrical actuation are usually called *solenoid valves*. A solenoid consists of a coil and an armature. When an electrical current is applied to the coil, it generates an electrical field. The field causes the armature to move, and as it is pulled into the magnetic field, it moves the spool.

AC/DC

Solenoids can use either AC or DC. An AC solenoid has a large current draw when it is first turned on (in-rush current) but has a low draw after that (holding current). If an AC sensor cannot complete the shift, it continues to draw high current and burn out. This can also happen if a solenoid is turned on and off at a high frequency.

A DC solenoid draws constant current when it is energized. It is designed to handle continuous and high current. This prevents burnout if the shifting is incomplete or high cycle rates are required. DC solenoids generally operate at lower voltages and so are usually safer.

DESIGN PRACTICE

Let's complete the simple application we started earlier. We chose a 10-inch minimum stroke and a 8-inch squared area. Next we need to design the rest of the application.

We need to control the movement of the cylinder in both directions so we use a double-acting cylinder. We need to control the speed of the cylinder on the retract stroke, so we will add a flow valve. To control speed, we need to control the fluid leaving the cylinder, so the flow valve must always be placed where the air will leave the cylinder on the controlled movement. We chose a 4/2 control valve to switch the direction of the cylinder. The valve we chose has one solenoid and a spring return. We also added a air service unit to clean the air and control the pressure (see Figure 11-25).

Actuators are labeled with integers 1, 2, 3, and so on. For example, the two cylinders in an application would be numbered 1 and 2 (see Figure 11-26). Control elements are numbered .1, .2, .3 and so on. The regulator after the air supply, for example, is numbered 0.1. A directional control valve controlling cylinder 1 is numbered 1.1. Elements between a control element and an actuator are labeled .01, .02, .03, and so on. For example, a flow valve between directional control valve 1.1 and actuator 1 is labeled 1.01 (see Figure 11-27).

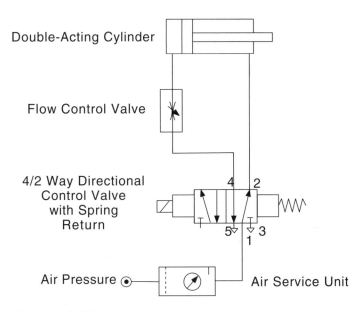

Figure 11–25 A simple double-acting cylinder application.

0	Air Supply Unit
1.0, 2.0, 3.0, etc.	Actuator
.1	Control Element
.01, .02, .03, etc.	Elements Between the Control Element and the Working Element

Figure 11–26 Numbering system for pneumatic components.

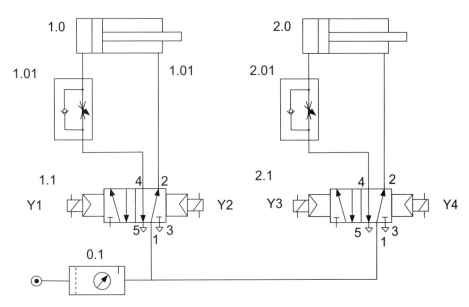

Figure 11–27 Numbering of components in a pneumatic schematic.

VACUUM

A vacuum is often used in industrial applications. When a small vacuum is needed, for example, to pick up small parts with a suction cup, a venturi vacuum generator system is often used. A venturi system uses positive air pressure to create a flow through the venturi, which then draws air in through the port attached to the suction cup(s). When more vacuum is required, a vacuum pump is used.

QUESTIONS

1. Label each component in the following diagram and describe its function. Then number each component according to the standard numbering scheme.

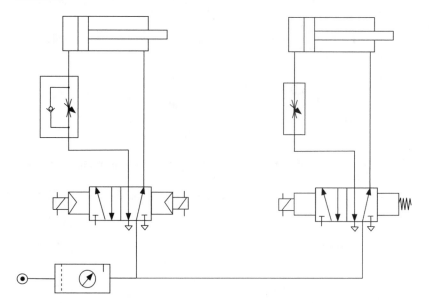

2. You are asked to design a pneumatic clamp for a system that must clamp with 750 pounds of pressure. You have 80 PSI available. What size cylinder should be used?

3. Draw the following valves and explain their operation:
    a. 3/2 solenoid valve with spring return
    b. 4/2 solenoid valve with spring return
    c. 4/2 solenoid valve with dual solenoid

4. Explain how to control the speed of a cylinder.

5. Design the following. We have a 300-pound load to move 5 inches. The speed must be controlled on the extend stroke. We have 100 pounds of

air pressure available. Calculate the cylinder size and choose the appropriate components and draw the circuit.

6. Draw and describe a double-acting cylinder.

7. Draw and describe the operation of a single-acting, spring return cylinder.

8. Describe the operation of a 4/2 valve.

9. Describe the operation of a 5/2 valve.

10. Draw a circuit with an appropriate valve to control a double-acting cylinder.

chapter

Fundamentals of Process Control

12

This chapter covers the fundamentals of process control. Open and closed loop systems will be examined. Proportional integral and derivative control will all be covered.

OBJECTIVES

Upon completion of this chapter you will be able to:
1. Define these terms: *open* and *closed loop, feedback, summing junction, command, error, disturbances,* and *damping*.
2. Describe the principle of on/off control.
3. Describe the principle of proportional, integral, and derivative (PID) control.
4. Describe tuning a PID system.

CONTROL SYSTEMS

Control system performance can be evaluated in many ways. The most important characteristic is stability, which indicates that a system can control smoothly without undue oscillation or overcorrection. Another characteristic is how closely the system can control to the setpoint, also called the *setting* or *command*. The difference between the setpoint and the actual value of the variable is called *error*. The smaller the error is, the better the control system. The objective is zero error, but this is not actually possible because the system must always respond to disturbances to try to make the actual value equal the setpoint.

A third characteristic is *steady state*. A system must be designed with an allowable error. For example, an engineer might be asked to design a temperature control system to regulate temperature to a value of 375 degrees plus or minus 2 de-

249

grees. This means that the allowable deviation in the actual system is 373 degrees to 377 degrees. When the system is operating at a given setpoint under normal operating conditions it is called *steady state*.

Another important characteristic of a control system is how quickly it can respond to an error and correct it. In many systems, speed of response is more important than very close control to setpoint.

All of these factors are interrelated. An adjustment that attempts to improve the system's accuracy reduces its stability. An adjustment to make a system more responsive to an error reduces the stability. In process control, as in life, system adjustment often seeks to achieve a happy medium.

Open- vs. Closed-Loop Control Systems

There are two basic types of process control: open and closed loop. Open loop is the simplest, so we examine it first. Figure 12-1 shows a level control system that has a tank with an output pipe and a valve that allows input to the tank. Figure 12-2 shows a block diagram of an open-loop system. Block diagrams are often used to illustrate control systems. The open-loop system in Figure 12-2 has no feedback since it has no sensor to sense the fluid level. Open-loop systems have no sensor, so they can provide no feedback.

Let's examine the simple level control application in more detail (see Figure 12-1). First imagine that the valve has only two positions; completely open or completely closed. Note that two lines have been drawn on the tank; a high-level indicator and a low-level indicator. If the level in the tank goes above the high-indication line, the operator should close the valve. If the level gets below the low-level indication line, the operator should open the valve. This is obviously very crude control but works if the operator stays awake and is attentive to the process. This is simple on/off control. The operator closes the loop, if the operator's attention is diverted, the level system is open loop and the tank will certainly overflow or empty itself.

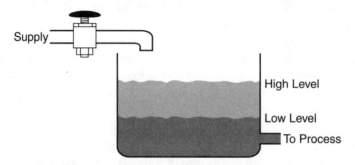

Figure 12–1 Level control system.

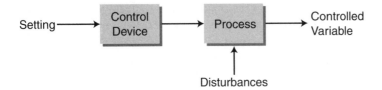

Figure 12–2   Block diagram of an open-loop control system.

Closed-Loop Control System

Closed-loop is a system that has feedback and can self-correct for various changes.

We can improve this system if we make it capable of self-correction. Let's use the typical home sump pump for our example (see Figure 12-3).

The float was added to provide feedback and to activate a switch that turns on the pump. The float can be adjusted up or down on the rod to change the setpoint. In our home sump pump, the setpoint is the level of water in the sump hole.

As with other on-off systems, the level is never held right at setpoint. Under normal conditions, the water gradually rises until the float on the rod forces the pump switch on, and the level falls rapidly until the float rod forces the switch to turn off. The water then gradually rises again. Under abnormal conditions, such as during a large spring thunderstorm, the water rises very quickly in the sump hole, and the float rises and forces the pump switch on. If it is raining hard enough, the level might actually rise above the setpoint if the pump cannot keep

Figure 12–3   Typical home sump pump.

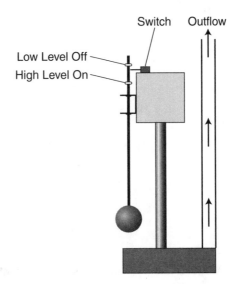

up, but the pump runs until the water level subsides. While the on-off home sump pump does not control set point (one exact level), it certainly does a more than adequate job of controlling the level of water.

Another example of a closed-loop on-off control system is the home furnace (see Figure 12-4). Note in the drawing that the home heating system is relatively simple. Its thermostat located on the first floor of the house measures the actual temperature in the house. The furnace is often found in the basement. If the temperature drops below the set point, the thermostat transmits a signal to the furnace, which opens the gas valve completely and turns on the fan to circulate heated air from a heat exchanger through the house. When the temperature rises above the set point, the thermostat opens, and the furnace valve is closed completely. The set point in this system is the desired operation point, perhaps 78 degrees.

Figure 12-5 is a block diagram of the home temperature control system. The process block always determines how the other blocks function. A feedback device must be chosen to measure the process variable associated with the process (the thermostat in this example). The adjusting device (furnace) must be chosen to adjust the manipulated variable (heated air) used to change the process variable (house temperature).

The homeowner sets the set point on the thermostat. The thermostat acts as a comparator, also called a *comparer or summing junction* (a very descriptive name for the comparer that accurately describes its function), that compares the set point

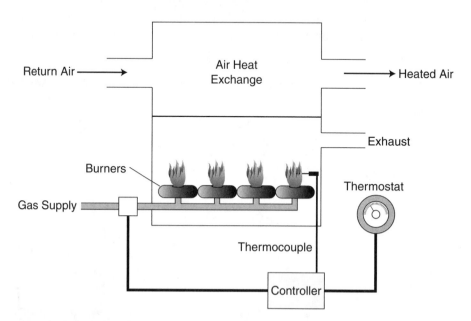

Figure 12–4 Closed-loop off-on control system for a furnace.

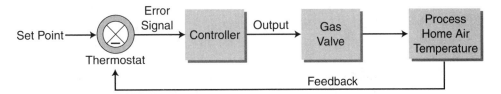

Figure 12–5   Block diagram of a home temperature control system.

and the air temperature in the room. The summing junction compares the set point and the feedback from the system and generates an error. The thermostat compares the set point and feedback (air temperature in the room) and generates an error signal to the furnace. In effect, the thermostat closes a contact to the furnace if the temperature is too low and opens the contact if the temperature is too high.

The summing junction has two inputs: the command (set point) and the feedback signal. The summing junction's output is an error signal used to control the final correction device. Note in Figure 12-5 the negative sign on the feedback connection. This means that the negative feedback principle is used. The polarity of the feedback signal is opposite polarity and is then summed with the command to generate an error signal.

Feedback   *Feedback* provides current information about a variable we are trying to control, such as level. Many types of sensors often called *transducers* can provide feedback. Transducers convert one form of energy to another form. In the usual home furnace, the thermostat's ambient air temperature is converted to a mechanical change in a bimetallic strip in the thermostat that closes or opens an electrical contact (see Figure 12-6). Most transducers for process control systems convert the measured variable (in the furnace, pressure) into an electrical signal that can be measured. In this case, the feedback device is the float, which changes

Figure 12–6   Principle of a thermostat.

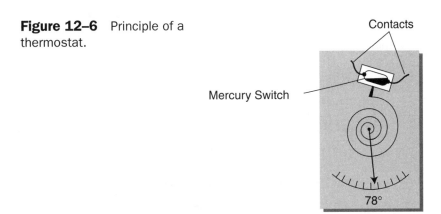

level into mechanical movement to a change in valve position. In other cases, thermocouples convert temperature to proportional voltage and tachometers convert velocity (motion) to proportional voltage.

Controlled Variable   The *controlled variable* is what is actually manipulated to control the *process variable*. It is not always possible to directly alter the process variable. For example, consider a tank that has a steam line running through it (see Figure 12-7). An operator can control the temperature of fluid in the tank only by controlling how much steam flows through the steam line. In this example, the controlled variable is the *flow rate of steam*. The process variable is the *temperature of the fluid in the tank.*

The operator can improve the system's operation by installing a variable valve at the input to the tank. The valve could then be set so that the flow into the tank from the valve approximately equals the tank's flow out (outflow). If the outflow is constant, the operator can get the input valve position quite close to perfect and the level would stay pretty much in the middle between the upper limit and lower limit lines. If the level falls, the operator opens the valve slightly; if it rises, the operator closes the valve slightly.

Imagine that the operator set the valve perfectly so that the inflow exactly equaled the outflow. The operator observed the system for 30 minutes and was sure it was operating properly. The operator decided to take a long break. It would not take very long before the tank is empty or overflows. The system is open-loop the minute the operator leaves, meaning that there is no feedback. The operator is not there to watch the process. The next feedback would probably be the boss's comments to the worker when the tank overflowed or ran empty.

Disturbances   The operator thought the valve was set perfectly but the tank overflowed the minute he left for several reasons. The tank outflow cannot be perfectly stable; it varies. The inflow cannot be stable either. Think about the water pressure in your house. The pressure in the system is constantly varying. The

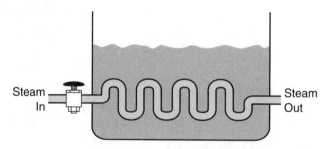

Figure 12–7   Tank with a steam line for heating.

washer might be running, someone may be taking a shower, someone could be watering the lawn, or the neighbor is washing the car. These are all disturbances to our system; every system always has disturbances.

A *disturbance* is anything that disturbs or disrupts the process. Consider the tank level example. There would be many possible disturbances. Imagine that the tank just contains water. The inlet pipe comes directly from the city water department. The water pressure varies continuously, depending on current usage throughout the city. There is a story that city water pressure is always the lowest during half time of a super bowl. Why would that be?

Low water pressure is an immediate disturbance to the system. Less fluid would be flowing in, which causes the level to drop in the tank example. When the second half of the super bowl begins, no one is using the water, and the water pressure goes up. Now too much water enters, and the level increases. So water pressure is continuously varying. It is a constant disturbance to the system.

The outflow also varies depending on the process using the water from the tank. Temperature can also have an effect on the viscosity of fluids, which affects flow rates. Imagine a thick fluid, like molasses. Cold temperatures make the fluid thicker and slow its flow rate dramatically. If the molasses is hot, the flow rate increases dramatically. Another reason could be that over time the inside of the pipes have corroded or built up scale, which impedes the flow and disturbs the system.

The home furnace also has many disturbances including the air temperature outside; the wind; leakage through windows, walls, and doors; and someone opening a door.

Systems experience many disturbances, and affects the process. Some have a dramatic effect on the process and some have only a small effect. A good control system is able to correct for all disturbances.

Consider the home furnace. The homeowner chooses a setpoint of 78 degrees. The furnace attempts to keep the temperature in the house at 78 degrees. The home furnace has an on-off controller. If the temperature is above 78, the furnace is shut off. If the temperature is below 78 degrees, the furnace is turned on (see Figure 12-8). The gas valve in a home furnace is either completely open (100 percent open) or completely closed (0 percent open).

Figure 12–8 Furnace on-off vs. setpoint.

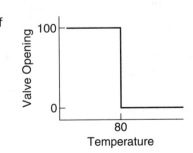

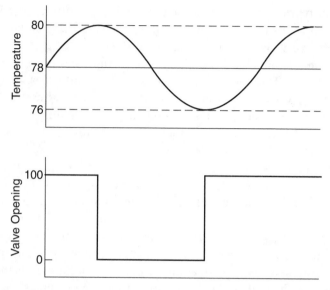

Figure 12–9 Graph comparing the output to the response for a simplified on-off controller.

A graph of the temperature in the home would be a sine wave (see Figure 12-9). When the furnace is on, the temperature graph is rising; when the furnace is off, the temperature graph is falling. The graph of the furnace response is a square wave however, because it is either on or off.

In reality, the furnace cannot change states at exactly one temperature. If the furnace had to change from on to off, or off to on at exactly 78 degrees, it could not operate. There needs to be a narrow temperature band where the furnace does not change states. This is typically called *deadband* or *hysteresis*.

To help illustrate the concept of deadband, consider a train crossing. Imagine two sensors, one on the left of the train crossing and one to the right of the train crossing. If a train crosses either of these sensors, the crossing light should flash its warning and the gates should close after a delay. The gates should stay closed until after the train completely leaves the crossing. Could the gates open after the first sensor does not sense the train anymore? No! The train may have left the first sensor but stopped in the crossing. Could the gates wait to open until after the first sensor and the second sensor no longer sense a train? This would not be safe either. It would work for a train that is longer than the distance between the two sensors, but what would happen with a short train? This situation would not be safe either because the first sensor would turn on and then off as the train passed by it but the train is too short and the second sensor would not be on, so the gates open. Logic in this case becomes very important. It ensures that the first sensor is on before the gates are closed. The logic then watches the second sensor to be sure that it turned on and off before the gates were opened (note that this logic is not perfect either). This example does help explain deadband. We do not want any change in the gate position while the train is between the sensors. We do not want the gates to reopen until the train has gone completely past the two sensors.

In the simple temperature control system, we do not want the furnace to change state unless the measured temperature changes completely through the deadband. Study Figure 12-10. Notice the 4-degree deadband. If the temperature is below 76, the furnace will turn on and stay on until the temperature exceeds 80 degrees. The furnace will then turn off when the temperature exceeds 80 and will remain off until the temperature falls below 76 degrees. Note that the furnace never changes state in the deadband. It changes state only if the temperature is above or below the deadband. As a matter of fact, the valve will never change state unless the measured value (temperature) goes completely through the deadband. In other words, if the temperature is below 76, the furnace turns on and the temperature begins to rise. The temperature moves into the deadband area, and the furnace remains on. It stays on until the temperature rises above the deadband (80 degrees). Now the furnace will remain off until the temperature falls all of the way through the deadband and below 76 degrees.

Note that an on-off controller can never maintain the actual temperature at exactly the setpoint. It continually oscillates above or below the setpoint. The oscillation frequency changes depending on process conditions. Consider your house on a nice fall day with the warm sun shining through the windows. The furnace does not run very often. The temperature oscillation in the house is very slow. Maybe the furnace only has to run once every hour. Now imagine a cold day in

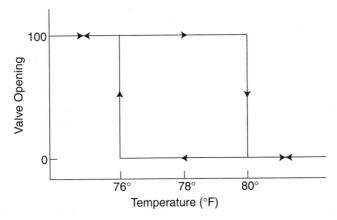

Figure 12–10 Comparison of the output to the response for an on-off controller with hysteresis.

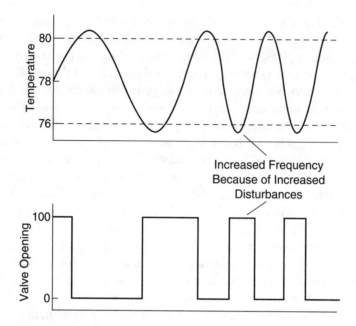

Figure 12–11 Graphs of on-off control.

February with the wind blowing hard outside. The furnace is running every five minutes to try to keep the temperature at 78 degrees. In this case, it operates at a much higher frequency (see Figure 12-11). Note that the amplitude is approximately the same but the temperature oscillates at a much higher frequency.

Imagine a normal December day in Wisconsin (10 degrees or so). Imagine the furnace is running about once every 30 minutes. What would happen if we narrow the deadband to 2 degrees instead of 4? The furnace would turn on and off more frequently to maintain the closer temperature. What if we widened the deadband to 8 degrees? The furnace would not have to run very often but the temperature amplitude (variation in temperature) would dramatically increase, and our comfort level would decrease.

Thus, there are limits to increasing the deadband. If we widen it too much, the furnace will not run very often but will run longer but we will not be comfortable. There is, however, a limit to how much we can narrow the deadband because at some point, the system will not be able to keep up with the rapid on and off commands.

Advantages and Disadvantages of On-Off (Closed-Loop) Control On-off control is inexpensive and simple to implement and maintain. It is more than adequate for many control applications.

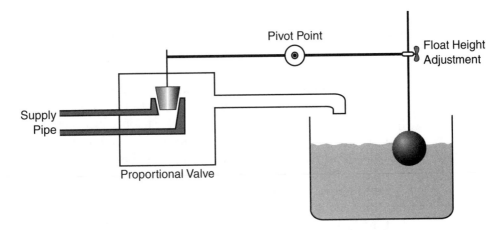

Figure 12–12 Closed-loop level control system.

An on-off controller is not able to keep the temperature at the exact set point. On-off controllers have permanent oscillation. The temperature always oscillates above and below the set point. An on-off controller also needs to have a deadband. This means that on-off controllers cannot hold tight control of the variable (temperature in this example).

Proportional (Closed-Loop) Control

Figure 12-12 shows a closed-loop level control system that is similar to the typical basement sump pump and that has feedback. The block diagram for it is shown in Figure 12-13. The float measures the height of the fluid in the tank and provides the feedback. The float is attached to a rod that automatically opens or closes the input flow valve. The valve can adjust between 0 percent and 100 percent open. Study the pivot point for the float rod; it can be adjusted to the right or left. This controls how much effect the float has on the valve position. If the pivot is moved to the right, the float has a large effect on valve position. Even a small

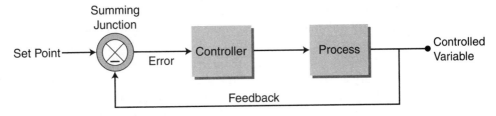

Figure 12–13 Block diagram.

movement of the float dramatically changes the valve position. If the pivot point is adjusted to the left, the system is less responsive. A change in the float height has less effect on valve position. The fluid level (setpoint) is adjusted by moving the float up or down the vertical rod.

Consider the operation of this system. If the outflow from the tank is relatively constant, we should find a point for the pivot that allows the system to operate quite nicely. The float would cause the valve to open just enough to allow approximately the same amount of fluid into the tank as is flowing out. If we could find the perfect point for locating the pivot, the inflow would always equal the outflow. In reality, there is no perfect pivot point.

If the level varies, it must be corrected. For example, consider the tank level system. If the commanded level in the tank is 5 feet of fluid and the feedback from the float shows 4 feet, there is a *one-foot error* (5 feet − 4 feet). The *summing junction* generates an error and then changes the valve position (generate an error signal) to correct for the one-foot error. In this case it causes the valve to open more and enable more flow into the tank. This correction continues until the setting (commanded level) equals the feedback and the summing junction generates a zero error.

Overadjustment   Now imagine the pivot moved from the "equilibrium" point to a point farther to the right. Now any movement in the float would cause a dramatic change in the valve position. Assume that the fluid level decreased slightly. The float would move down a small amount but cause a drastic change in valve position. The valve in this case would open a great deal and allow too much fluid to flow into the tank. The level would increase rapidly and the float would move up. The float would cause the valve to close too much and the inflow would be too small to keep up with the outflow. With the valve too far to the right, the response to any error would be too great and the system would begin to oscillate. In other words, it would start to fight itself to try and make rapid changes to keep up with its overcorrection. The control in this case is poor and the level would fluctuate rapidly above and below the set point.

Now imagine that we move the pivot point too far to the left. A change in the float would cause very little change in the valve opening. Assume that the outflow increases. The float begins to drop, which opens the valve slightly. The valve does not open enough so the level continues to drop, which drops the float and opens the valve more but not enough. The level will continue to fall until the float has opened the valve enough to make inflow equal to outflow. With the pivot in this position, the response will be very slow, and the height of the fluid in the tank will vary a great deal. It would be very slow to correct for a change in level.

The proper size valve to control the flow is important. If a chosen valve is too small, the valve might not be able to correct for certain errors. For example, the tank would eventually empty if the outflow was greater than the inflow and the valve could not allow enough flow to keep up. If the valve is too large, it could have a tendency to overcorrect and oscillate on and off more. The size of the valve and the setting (pivot position in this simple application) are crucial to proper control of a system. A valve should generally be chosen so that it operates between 30 percent and 70 percent open during normal operation.

Stability   *Stability* is a key in any system. A system can be very stable if the pivot point (proportional band) is set correctly. If we set the proportional band too small, the system overcorrects for any error and becomes unstable. The level would vary a great deal, and the valve would be opening and closing erratically to try to maintain the level.

In a stable system the proportional band is set so that the valve makes appropriate adjustments so that the level remains relatively constant. Stability is one of the most important characteristics in any system. To be stable the system components must be appropriately sized for the application and the system must be correctly adjusted (tuned).

The proportional gain portion of *proportional, integral, and derivative* (PID) control looks at the magnitude of the error. The proportional response to an error has the largest effect on the system. Proportional control reacts proportionally to errors. A large error receives a large response. (In the case of a large temperature error, the fuel valve would be opened a great deal.) A small error receives a small response.

Imagine a furnace that can be heated to 1500 degrees. There is a portion of this range of temperature at which the response of the system is proportional. For example, let's assume that between 1000 and 1500 degrees, the system adjusts the valve opening in proportion to the error (see Figure 12-14). Below 1000 degrees, the valve is open 100 percent. Above 1500 degrees, the valve is open 0 percent. The *proportional band* in this case is 500 degrees. Proportional band is normally given as a percentage (see Figure 12-15). The percentage is calculated by dividing the proportional band in degrees by the full controller range and multiplying

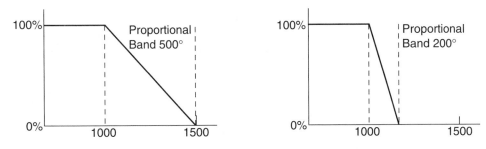

Figure 12–14   Graphs of proportional band between 1000 and 1500 degrees.

Figure 12–15 Proportional band example.

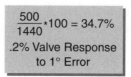

$$\frac{500}{1440} * 100 = 34.7\%$$

.2% Valve Response to 1° Error

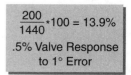

$$\frac{200}{1440} * 100 = 13.9\%$$

.5% Valve Response to 1° Error

by 100. The full controller range is simply the range of temperatures that the furnace can control. In this case let's assume that the full controller range is 1440 (1500 − 60). Are you wondering why 60 was subtracted? Assume that room temperature is 60 degrees. The furnace cannot control below room temperature, so that is excluded in our calculation. In this case the proportional band, in degrees, equals (500/1440) * 100, or 34.7 percent. The technician can adjust the width of the proportional band to make the system more or less responsive to an error. The narrower the proportional band, the greater the response to a given error (see Figure 12-16).

Proportional Control Gain   The gain of a system can be calculated as shown in Figure 12-17. For example, in the previous example, the proportional gain was 34.7 percent. If we divide 100 percent by 34.7 percent, we get a gain of approximately 3.

Figure 12-18 is an example of a disturbance on a system. The system's response to three different proportional bands is shown. The setpoint and the disturbance were the same in each case. The narrowest proportional band (20 percent) had the greatest response to the disturbance (error). The 50 percent proportional band had a smaller response to the same disturbance. The 200 percent proportional band had the least response to the disturbance. This simple illustration presents several important concepts. Note that none of the examples returned the actual temperature to the setpoint, but each did return the temperature to an equilibrium. Note also that the narrower the proportional band (most response to an error), the closer the system returns the temperature to the setpoint. The wider

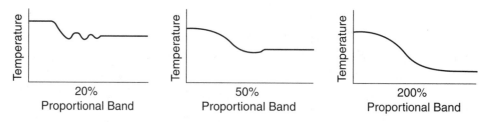

Figure 12–16 Figure showing the effects of a narrow and wide proportional band.

Figure 12–17 Proportional gain calculation.

$$\text{Proportional Gain} = \frac{100\%}{\text{Proportional Bond}}$$

$$\text{Gain} \sim 3 = \frac{100\%}{34.7\%}$$

the proportional band (least response to an error), the farther from the setpoint the new equilibrium is. Let's examine the reason for this. You may have to read this paragraph several times to understand this concept.

Assume that the furnace had been controlling the furnace on setpoint (assume 1000 degrees) at 60 percent valve opening. In other words, a 60 percent valve opening keeps the furnace's actual temperature at 1000 degrees. Now imagine that a disturbance occurs. The furnace operator does not completely close the furnace door, and the temperature drops, which increases the error. The controller opens the valve to 70 percent to bring the actual temperature back to the setpoint. The error begins to decrease as the temperature approaches the setpoint. But if the temperature gets back to the setpoint, the error is zero, the valve would return to 60 percent valve opening, and the temperature falls again. A proportional system can never return to setpoint. Some small error must be maintained to keep the valve at a slightly different position than the setpoint calls for.

Refer again to Figure 12-18. Note that the narrow proportional band was able to recover more fully from a disturbance because a narrow proportional band causes a larger response to an error. In other words, a 20 percent proportional band would respond 10 times as much to the same error as our 200 percent proportional band.

Note the difference in the graphs again. The narrow proportional band response shows some oscillation around the setpoint until the system settles. The second shows a smoother response and just a slight overshoot. The third example

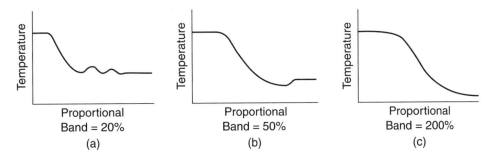

Figure 12–18 Effects of width of proportional band on response to a disturbance.

takes the longest to correct but is the smoothest response. Which is best? It depends on the application. If we need very quick correction for errors, the narrow band is best. If smooth response is crucial, the second or third band might be best.

Thus far we have assumed that a proportional controller could control on set-point if there were no disturbances. Unfortunately, this is not true. In reality and assuming no disturbances, there is only one point at which a proportional controller could control at exactly the set point.

Offset   If a proportional-only control system had its set point set exactly at the 50 percent valve opening needed to maintain the temperature, it could control right on set point, assuming no load changes or disturbances. This, however, is never the case. Proportional-only controllers can never control right on set point. The difference between the set point and the actual measured value is called *offset* (see Figure 12-19). The offset may cause the measured value to be below or above the set point. The amount of offset varies by how close the set point is to the 50 percent valve opening (ideal set point). The farther the set point is from the ideal set point, the larger the offset is (see Figure 12-20). Offset can also be caused by load changes. Remember that there is one ideal temperature assuming a given load. The valve would be 50 percent open to maintain that value. Now assume a larger load but the same set point. A 50 percent valve opening would not be able to maintain the set point anymore because of the larger load. So the actual value could not be held at the exact set point because if there was a zero error, the valve would return to 50 percent open and the actual value would drop again. Offset can be reduced by narrowing the proportional band. Remember that a narrow proportional band responds more to even a small error. Remember also that if the proportional band is narrowed too much, the system will become unstable. To illustrate this, imagine a proportional band of zero. It would be an on-off controller without hysteresis.

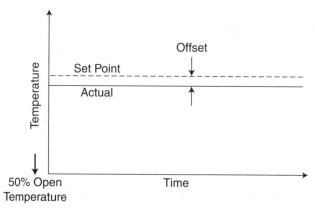

Figure 12–19   Offset in a proportional system.

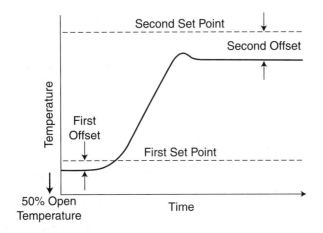

Figure 12–20 The farther the actual set point is from the ideal set point, the larger the offset.

Effect of Load The load on a system affects how the system responds. A heavier load requires more valve opening to keep the system on setpoint. Figure 12-21 shows a system with three different loads. Imagine a furnace in which a small metal part needs to be brought up to temperature. Then imagine how the same furnace would have to respond to a large piece of metal.

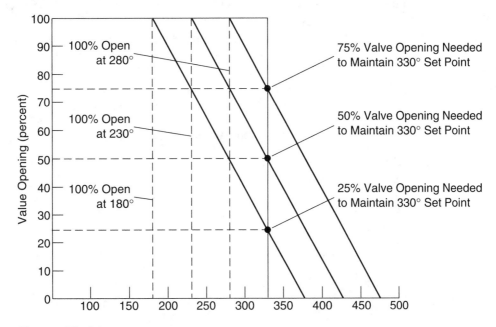

Figure 12–21 Effect of load on a system.

Figure 12–22 Industrial furnace.

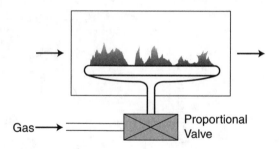

Let's examine proportional gain first. The proportional factor responds to the magnitude of an error: large error—large response, small error—small response. Proportional control means we make the response to an error proportional to the error. Proportional control can help eliminate some of the shortcomings of on-off control.

Consider the simple temperature control system in Figure 12-22. This system represents an industrial furnace with a proportional valve to control the flow of gas to the burner. The valve can be controlled between 0 percent and 100 percent open. To keep the numbers simple, assume the furnace can control the temperature between 70 and 1070 degrees, or 1000 degrees. This means that the valve is 100 percent open at 70 and 0 percent open at 1070. Thus 100 percent of the valve opening is spread over 1000 degrees. The graph in Figure 12-23 shows the relationship between valve opening and temperature for this system. For this graph assume that the 50 percent valve opening represents the setpoint. If the temperature rises above 570, the valve must close proportionally to bring the temperature back

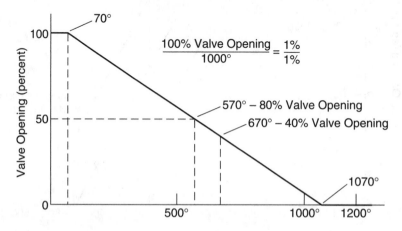

Figure 12–23 Relationship between valve opening and temperature.

to 570 degrees. For any 10 degree temperature change, the valve changes 1 percent (see Figure 12-23). Thus, if the temperature rose to 580 degrees, the valve would close 1 percent more to bring the temperature back to 570 degrees. If the temperature fell to 560 degrees, the valve would open 1 percent more. This system would have very low response to an error. For example, if the operator changes the set point from 300 to 350, the valve would change only 5 percent. Imagine that it takes five minutes for the furnace to get to the new temperature.

The highest temperature we need for our process is 450. What if we could change our system so that the valve would be closed at 70 degrees and 100 percent open at 570 degrees? This would improve the system response. In other words, the system would react more quickly to errors or to setpoint changes. This change can be done by revising our proportional band over which the valve changes from 0 to 100 percent open.

In the previous example, it took five minutes to change the temperature 50 degrees when the setpoint was changed. There was a 1 percent change in valve position for every 10-degree error. With the new smaller proportional band, the system would have twice the response to a given error. If the setpoint is changed from 300 to 350 degrees, there would be a 50 degree error. The system would now respond with a 2 percent valve change, twice the response. It could correct for a 50-degree error in approximately 2.5 minutes—twice as fast as the previous system.

Summary of Proportional Control   Proportional control acts on the magnitude of the error. It can eliminate the permanent oscillation that exists with an on-off controller. Proportional control also comes closer to setpoint than an on-off controller. A few problems are associated with proportional control, however. They cannot control exactly on setpoint because of permanent error and offset, and is appropriate for most slowly changing systems.

Integral Gain Control

Proportional control cannot solve all problems. Imagine the cruise control system in a car. We set the setpoint at 65 miles per hour. The cruise control sets the fuel flow rate. If there are no disturbances, the car would move at 65 miles per hour. There are many disturbances, however, such as going up a gradual incline. The speed would fall slightly based on the proportional fuel flow rate.

Proportional cannot correct for very small errors, called *offset* or *steady-state errors*. An example of a steady-state error might be driving your car on a level road with the cruise control on. The car does not maintain the exact speed that you chose. The proportional control is quite happy, however. The proportional control cannot correct small variations, or steady-state error.

The integral gain portion corrects for the small error (offset) that proportional cannot. Integral gain control looks at the error over time. It increases the importance of even a small error over time. Integral is determined by multiplying the error by the time the error has persisted. A small error at time zero has zero importance. A small error at time 10 has an importance of $10 \times$ error. In this way integral gain increases the response of the system to a given error over time until the problem is corrected. Integral can also be adjusted. The integral adjustment is called *reset rate*, which is a time factor. The shorter the reset rate, the quicker the correction of an error (see Figure 12-24). In hardware-based systems, a *potentiometer* makes the adjustment. The potentiometer essentially adjusts the time constant of a resistor–capacitor (*RC*) circuit. Too short a reset rate can cause erratic performance.

Proportional vs. Integral Gain Control   Proportional considers the magnitude of the error, but it is not able to control exactly on setpoint. There is always a small permanent error and offset in a proportional controller. Integral gain control corrects these small errors over time. The addition of integral control enables a proportional system to control on setpoint and can eliminate permanent error and offset.

Derivative Gain Control

Both proportional and integral have problems. Neither considers the rate of change in the error. A system should react differently to a rapidly changing error than a slowly changing error. Derivative gain control, sometimes called *rate time*,

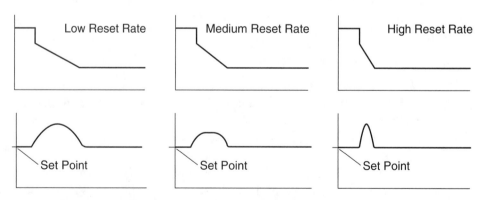

Figure 12–24   Effects of short, medium, and long reset rate. The vertical line in the three top graphs shows the proportional response to the disturbance. The proportional response to the error was the same in each example.

considers the rate of change in the error and attempts to look ahead and prevent overshoot and undershoot of a proportional controller. This is often called *damping*; it allows the gain (or proportional factor) to be set higher than would be possible without some derivative.

Consider the cruise control again. The car is going along at 65 miles an hour and all of a sudden begins going down a very steep hill. The proportional and integral controls were keeping the speed constant until now. The rate for change in the error is large now (the speed is rapidly increasing). Derivative control acts and makes a large change in fuel flow. This brings the speed back close to setpoint again so that integral may again act if a small error persists over time.

Let's consider another automobile example. A major problem with proportional control is that it cannot adjust its output based on the rate of change in the error (remember that the proportional term looks at magnitude of error, not rate of change in the error). We are driving on the highway and the brake lights on the car ahead of us come on. We react normally with what we think is appropriate brake pressure, perhaps 50 percent. This is the proportional response. Then we would watch the gap between our car and the car ahead of us. If the gap remains constant (constant error), our proportional response was correct and no additional response (derivative) is needed. If, however, the gap is decreasing rapidly (rapid rate of change in the error signal), we quickly apply additional brake pressure—a derivative response. Likewise, if the gap is increasing rapidly (large rate of change in the error signal), we reduce pressure on the brake pedal.

In other words, proportional responds first and then the derivative monitors the rate of change in the error. The derivative acts when a change in the rate of error occurs. We can control how quickly derivative responds to a change in the rate of error. Derivative causes a greater system response to a rapid rate of change than to a small rate of change. Think of it this way: If a system's error continues to increase, the control device must not be responding with enough correction. Derivative control senses this rate of change in the error and causes a greater response.

The derivative gain helps "damp" a system. If the goal is to have a system that will be very responsive and quickly correct errors, it is necessary to have a relatively high proportional gain. A high proportional gain causes the system to overshoot. The derivative gain can help damp the overshoot. When a system overshoots the commanded value (position, velocity, or temperature) the error rapidly increases.

The derivative gain responds to this rapid rate of change in the error and damps the proportional response. The derivative term provides a much higher proportional gain and a stable system. The derivative term also adds "stiffness" to a system. If we were to turn the shaft on a small motor with proportional but no derivative term, it would move relatively easily. When we add derivative gain, the shaft becomes much more difficult to move. The system fights more to maintain position or zero velocity.

Proportional, Integral, and Derivative Control   Proportional gain control considers the magnitude of the error. Integral control considers small errors over time. Derivative control considers the rate of change in the error. It can help damp a system. The addition of derivative allows the proportional gain to be higher and improves the response and stability of a system.

PID Control Systems

Proportional, integral, and derivative controls are used to control processes. PLC manufacturers have approached PID control in two ways: Some offer special-purpose processing modules, and some utilize standard analog/digital I/O with PID software in the CPU (see Figure 12-25). PID can be used to control physical variables such as temperature, pressure, concentration, and moisture content. It is widely used in industrial control to achieve accurate control under a wide variety of process conditions. Although PID often seems complex, it need not be. It is essentially an equation that the controller uses to evaluate the controlled variable (Figure 12-26). The controlled variable (temperature, for example) is measured and feedback is generated. The control device receives this feedback. The control device compares the feedback to the setpoint and generates an error signal. The error is examined by proportional, integral, and derivative methodology. Each of the three factors can be thought of as a *gain*. Each can affect the amount of response to a given error. We can control how much of an effect each has. The controller then uses these gains (proportional gain, integral gain, and

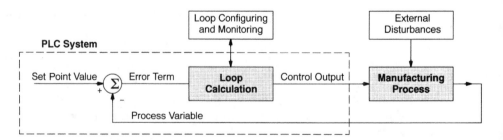

Figure 12–25   Block diagram of a process controlled by a PID controller. The system is a tank that controls temperature. A PLC controls the system. The CPU takes input from the analog input module and performs the PID calculation. The CPU generates an error signal and sends it to the output module (digital or analog). The output controls the fuel valve. (*Courtesy PLC Direct by Koyo.*)

Co = the output of the equation
K = the overall controller gain
1/Ti = reset gain (integral factor)
KD = rate gain (derivative factor)
dt = time between samples
bias = output bias
E = system error equal to the set point minus the
measured value.
E(n-1) = Error of the last sample.

Figure 12–26   PID equation.

derivative gain) to calculate a command (output signal) to correct for any measured error.

Figure 12-27 shows how a PLC implements the PID algorithm. Note the loop table in the diagram; it holds the user parameters for each gain.

Rockwell Automation PID Instruction   Rockwell Automation has a PID instruction that can control a closed loop using inputs from an analog input module and providing an output to an analog output module. The instruction allows the user to convert an analog output to a time proportioning on-off output for driving a heater or cooling unit. A Rockwell Automation PID instruction is shown in Figure 12-28.

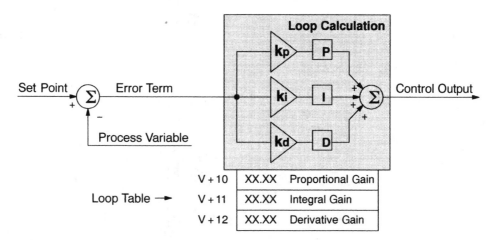

Figure 12–27   Block diagram of the PID loop calculation. (*Courtesy PLC Direct by Koyo.*)

Figure 12–28   Rockwell
Automation PID instruction.

```
┌─ PID ─────────────────────────────┐
│                                    │
──│  PID                              │
  │  CONTROL BLOCK          N10:0     │
  │  PROCESS VARIABLE       N10:28    │
  │  CONTROL VARIABLE       N10:29    │
  │  CONTROL BLOCK LENGTH   23        │
  └────────────────────────────────┘
```

The control block is a file that stores the data required to operate the instruction. The file length is fixed at 23 words. The user enters the address of the first integer word to be used. For example, if the user enters N7:10, the instruction allocates words N7:10 through N7:22.

The process variable (PV) is an element that stores the process input value. This address normally is the address of the analog input word. This value could also be the address of an integer value if the user chooses to prescale the input value to the range of 0–16383.

The control variable (CV) is the address that stores the output of the PID instruction. The output ranges from 0–16383; 16,383 is 100 percent of the value. The user can use the address of the analog output word or an integer that could then be scaled to the particular range needed.

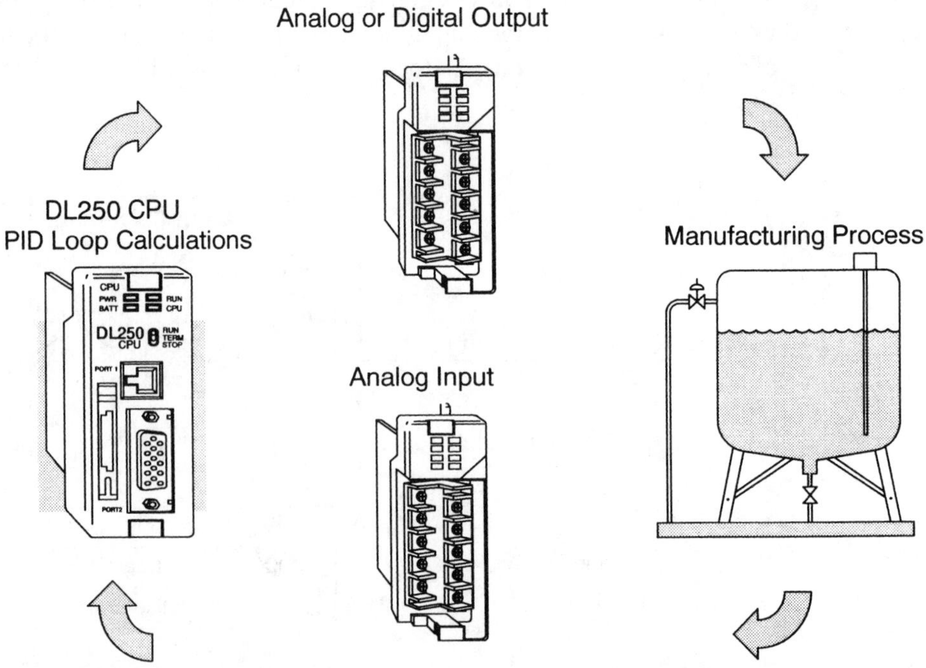

Analog or Digital Output

DL250 CPU
PID Loop Calculations

Manufacturing Process

Analog Input

Figure 12–29   PID-controlled tank process. (*Courtesy PLC Direct by Koyo.*)

After these addresses have been filled in, the user double clicks on the instruction and configures the parameters that determine how the instruction will perform for the application.

Most larger PLCs offer PID capability through the use of software and analog I/O modules. Some offer the capability through special PID modules. The PLC Direct 250 CPU offers extensive PID capability. Figure 12-29 is an example of the use of a PLC to control a tank process using PID control.

Note that there is an analog input module so that the process can be monitored. The output can be either analog or digital. In this application, a PLC Direct 250 processor was used. The 250 DirectSOFT programming software allows the user to use dialog boxes to create a formlike editor to set up the loops. DirectSOFT's PID Trend View can then be used to view and tune each PID loop.

Figure 12-30 is a diagram of the system. The process in this case is the level of fluid in the tank. Note the external disturbances that affect the process, which are always present. Temperature changes can cause the viscosity of the fluid to change,

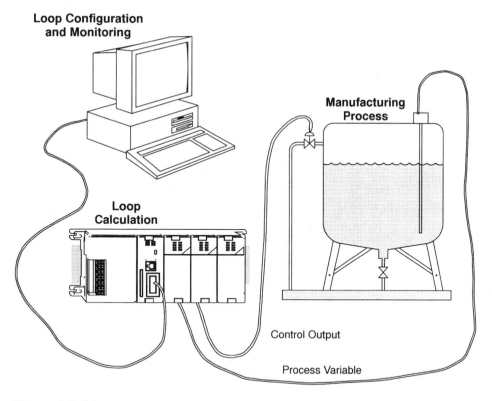

Figure 12–30   PLC system to control the tank process. (*Courtesy PLC Direct by Koyo.*)

the pressure of the supplied fluid into the tank can vary, and so on. There are many disturbances, and their individual effects are constantly changing. The PID control system must overcome these.

Feed Forward   Feed forward is an enhancement to PID control that improves system response when a predictable error occurs from changing set points or commands. For example, after adjusting the PID parameters as well as is possible, a predictable following error exists when a new position command is sent. Feed forward can be used to compensate for these errors. Figure 12-31 shows a block diagram of *feed forward* or *bias*. The feed forward term is "fed forward" around the PID equation and summed with its output. Without feed forward when a new command is issued, the loop does not know what the new operating point is. The loop essentially must increment/decrement its way until the error disappears. When the error disappears, the loop has found the new operating point. If the error is somewhat predictable (known from previous testing) when a new command is issued, we can change the output directly using feed forward. This term can be added in many controllers to help improve system response. If used correctly, it can also help reduce integral gain and improve system stability.

Tuning a PID System   Tuning a PID system involves adjusting potentiometers or software parameters in the controller. Many systems now employ software-based PID control allowing the technician to change a software parameter value for each factor. The goal of tuning the system is to adjust the gains so that the loop will have optimal performance under dynamic conditions. Tuning is often a trial and error process, but formulas can be used to develop approximate PID settings. One of these will be examined later in this chapter.

The technician should follow the procedures in the technical manual for the system being adjusted and must take care whenever making changes to a system. The following example of tuning a drive is based on a system in which the techni-

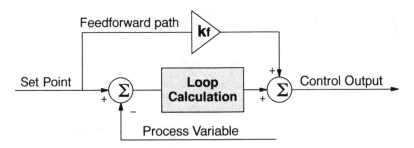

Figure 12–31   Feed forward compensation. *Courtesy PLC Direct by Koyo.*

cian can watch the actual error terms while the system is operating. In general, the tuning procedure occurs as follows:

1. Turn all gains to zero.
2. Adjust the proportional gain until the system begins to oscillate.
3. Reduce the proportional gain until the oscillation stops and then reduce it by about 20 percent more.
4. Increase the derivative term to improve the system stability.
5. Next raise the integral term until the system reaches the point of instability and reduce it slightly.

An oscilloscope can be used to view system performance during tuning. In general, a command signal, often a square wave, is issued to the drive. By comparing the square wave command on the oscilloscope screen to the system response, the technician can make adjustments to the PID parameters. The system response can be traced by checking the tachometer or the current feedback.

Zigler-Nichols PID Tuning   A well-tuned control loop should exhibit the following characteristics: stability, acceptable performance at steady state, and adequate response to setpoint changes and disturbances.

One of the oldest and most widely used PID tuning techniques is called the *Zigler-Nichols method*. The user makes adjustments to a closed loop system until steady oscillations begin to occur. The user must find the proportional gain (critical gain) that causes a steady state oscillation to occur and then measure the period (critical period) of the oscillations. The user then calculates settings for proportional, integral, and derivative settings based on the two critical gains and the critical period that caused the steady oscillation.

The Zigler-Nichols method begins by allowing only the proportional factor to influence the system. The integral and derivative are set so that they have zero effect on the loop. Next the user gradually increases the proportional gain to establish the point at which the system begins to steadily oscillate around the set point. This is the critical gain. The user must also create small disturbances during this phase of the tuning to force the system into oscillation. The user should then note the critical gain K_C when the measured variable first begins to oscillate around the set point. At this point the user must record the critical period (T_C) of the oscillations in minutes.

For a controller that utilizes only a proportional gain, the proportional gain is calculated using the following formula:

$$K_p = 0.5 * K_C$$

The values for a proportional plus integral controller could be calculated as follows:

$$K_p = 0.45 * K_C$$

$$T_I = T_C/1.2$$

If the quarter-amplitude criteria is desired, use this formula:

$$T_I = T_C$$

Note that the proportional gain must then be adjusted to achieve the quarter-amplitude response. The values for a full proportional, integral, and derivative control system are calculated using the following formulas:

$$K_p = 0.6 * K_C$$

$$T_I = T_C/2.0$$

$$T_D = T_C/8$$

Note that for a proportional only controller, the gain is found by multiplying the critical period by 0.5. When the derivative is added, the proportional gain is calculated by multiplying the critical period by 0.6. This means that the proportional gain can be set higher (approximately 20 percent) when a derivative is added.

If quarter-amplitude response is desired, the formulas are:

$$T_i = T_C/1.5$$

$$T_D = T_C/6$$

Quarter-amplitude tuning (see Figure 12-32) implies that the amplitude of each peak of a cyclic response must be one quarter of the preceding peak. The proportional gain must then be adjusted to achieve a quarter-amplitude response. Many combinations of PID values can achieve acceptable system performance.

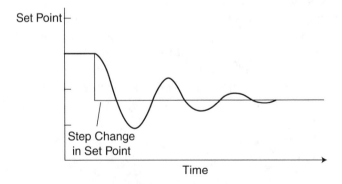

Figure 12–32   A step change in input and a quarter-amplitude response.

Tuning Problem   A technician needs to tune a temperature control loop for a new application. The technician begins adjusting the proportional band down from 100 percent while applying regular disturbances for the system to react to. When a 40 percent proportional band is reached, the system starts to oscillate. The technician measures the time period and finds it to be 8.5 minutes.

Tuning Problem Solution   The critical gain can be calculated from the proportional band using the following formula:

$$K_C(\text{critical gain}) = 100/\text{Proportional band}$$

$$2.5 = 100/40$$

The critical gain (K_C) is 2.5. Next the proportional gain is calculated:

$$K_p(\text{proportional gain}) = 0.6K_C$$

$$1.5 = 0.6 * 2.5$$

The proportional gain is 1.5.
   The setting for the integral is found using the following formula:

$$T_I = T_C/1.5$$

$$5.7 = 8.5/1.5$$

$$T_I = 5.7 \text{ minutes}$$

The setting for the derivative is found as follows:

$$T_D = T_C/8$$

$$1.06 = 8/8.5$$

$$T_D = 1.06 \text{ minutes}$$

Industrial Process Reaction Delay

Industrial processes do not fully respond to a change in input instantly. Imagine a tank of water that needs to be heated to a given temperature. The tank has a capacity to hold heat just like a capacitor has a capacity to hold a charge. The concept of time constants applies to both. Study Figure 12-33. It took 200 seconds for the fluid in this system to get to the new temperature. As you can see, it took about one time constant to get to 63 percent of the new setpoint. It takes five time constants to get to the new temperature. Remember that a capacitor charges to 63 percent of its capacity in one time constant also.
   The tank has thermal capacity but also thermal resistance. All materials have a natural reluctance to conduct heat. In this example the steam heats the walls of the

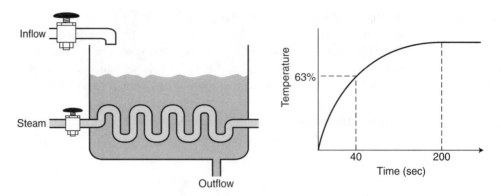

Figure 12–33 Time constant principle.

steam pipe, not the water directly, and the pipe heats the water. The thermal resistance depends on the thermal conductivity of the pipe wall, its thickness, and the surface area of the coils. The thermal capacity depends on the specific heat of the liquid and the size of the tank. The thermal time constant is based on the combination of the thermal resistance and the thermal capacity.

Almost all industrial processes exhibit process reaction delay (time constant delay). The length of the delay varies widely from very short times to hours, but virtually all processes experience a delay between taking corrective action and seeing the final result.

Transfer Lag Many processes have more than one resistance-capacity combination (see Figure 12-34). For example, gas is used in a furnace to create heat in the heat chamber. The heat must move through the wall of the heat chamber into the air in the chamber. The air then heats the parts in the furnace. If the parts could be heated directly instead of using this indirect method, the curve is a normal *RC* time

Figure 12–34 Furnace with transfer lag.

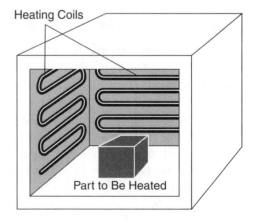

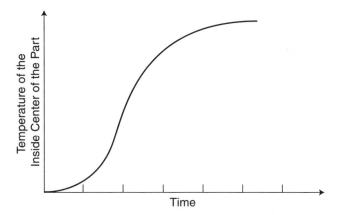

Figure 12–35 Graphed temperature versus time for a two-capacity process.

constant curve. But the air must be heated first and then the parts, resulting in two capacities: the capacity of the air and the capacity of the part as well as two resistances: the resistance of the wall of the heating chamber to transmit heat and the resistance of the part to transfer heat. The curve looks more like the one in Figure 12-35. The significance of transfer lag is that it makes control more difficult.

Transportation Lag Figure 12-36 is an example of transportation lag. The process involves heating a fluid in a tank to a given temperature and then sending it to the process. The temperature is measured at the process rather than in the tank. Opening the steam valve would produce an immediate response in the tank. Even though it might be small, an immediate change could be measured. Assume that the fluid takes five seconds to travel down the pipe to the process. This means that even though the change is immediate in the tank, it will be five seconds before the temperature sensor

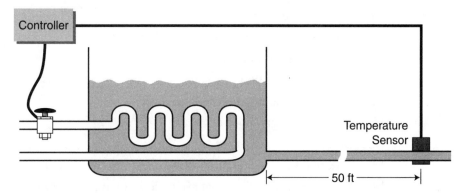

Figure 12–36 Transportation lag example.

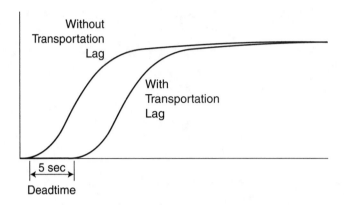

Figure 12–37 Response with and without transportation lag.

senses the change. This period between making a change and sensing it is called *deadtime*. Here it is five seconds. To put this in a more meaningful situation, imagine what would happen if the human brain could not respond to a change for five seconds. It would be impossible, for example, for a person to drive a car if it took five seconds to see the effect of moving the steering wheel. The graph in Figure 12-37 shows the response of the system with and without transportation lag.

QUESTIONS

1. List four characteristics of a "good" control system.
2. What are the major differences between open- and closed-loop control?
3. Is the typical home sump pump an open- or closed-loop system?
4. Define the term *controlled variable*.
5. Describe the term *proportional band*.
6. Describe the term *deadband*.
7. How does narrowing the deadband affect a system?
8. Describe proportional control.
9. What is offset? What causes it?
10. What effect does load have on a system?
11. Describe integral control.
12. Describe derivative control.
13. Describe PID tuning utilizing the Zigler-Nichols method.
14. What is transfer lag? What effect does it have on a control system?
15. What is transportation lag? What effect does it have on a control system?

chapter

Process Control Systems

13

This chapter examines various types of process control systems and instrumentation.

OBJECTIVES

Upon completion of this chapter you will be able to:
1. Describe various flow, level, density, and temperature measurement devices.
2. Describe the types of signals used in process measurement.
3. Describe common methods of measurement and control for flow, level, density, and temperature measurement systems.
4. Explain the purpose and importance of calibration.

FLOW CONTROL PROCESS

The flow rate of a liquid, gas, or solid (particles) often must be controlled in industry. Flow control is accomplished by measuring the force of the flow and then using a variable valve. Some devices measure the actual flow; others measure changes in pressure to approximate the flow rate. The velocity of fluid flow depends on the amount of pressure forcing the fluid through the pipe; it can be determined as follows.

Flow = Cross-sectional area of the pipe × Average velocity of the flow

Fluid Flow through Pipes

The three main types of fluid flow in a pipe are laminar uniform, laminar nonuniform, and turbulent (see Figure 13-1). Most fluid flow applications involve turbu-

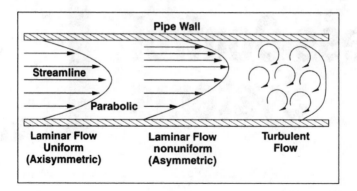

Figure 13–1 Three types of flow in a pipe. (*Figure published in the Rosemount Product Catalog, Publication No. 00805-0100-1025 © 1998, Rosemount, Inc. Used with permission.*)

lent flow, which occurs when velocities are high or viscosities are low. A value called a *Reynolds number* is used to represent the turbulence of flow. The density and viscosity of fluids as well as the friction created by the flow over pipe walls affect the flow rate. The pipe diameter and the viscosity of the fluid act as drag forces on the fluid.

In a given application, the specific gravity of the fluid and the pipe diameter remain constant, but the velocity and the viscosity vary. Note the parabolic shape and the flow in smooth layers (Figure 13-1). The highest velocity is at the center of the pipe, and the slowest is at the pipe wall because viscous properties of the fluid against the wall restrain it.

Flowmeters

Mass, positive displacement, differential pressure, and velocity are the four main types of meters available to measure flow. Selection of the proper flowmeter is crucial to the success of the application.

Mass Flowmeter One of the more interesting transducers used to measure mass rate of flow directly rather than the volume of the flow and changes in density is the Coriolis meter, which is one of the more commonly used measuring instruments. Its meters are very linear. It uses u-shaped tubes (see Figure 13-2). An electromechanical coil vibrates the tubes. When fluid passes through a u-shaped tube that is vibrated up and down, the flow causes the tube to twist. Motion sensors mounted on each side of the tube sense how much twist exists by measuring

Figure 13–2 Coriolis meter.

the differences in timing as the tube passes the sensor on each side. The Coriolis principle can be used to measure mass flow or density (Figure 13-3). The output from a Coriolis meter is a 4-20 mA signal that is proportional to the flow or the density of the fluid.

In process control 4-20 mA is a very common signal, and almost all devices (measurement and control) are available in 4-20 mA. There are several advantages in using 4-20 mA signals. A voltage signal is subject to losses as it travels through the wire; the longer the wire, the larger the loss. A current signal is not subject to this loss. Another advantage of a 4-20 mA signal is in tuning the system (calibration will be covered later). The "zero" adjustment for the system is adjusted until 4.0 mA is reached. If you read 3.98 mA you would adjust the "zero" back to 4.0 mA. If 0 was used instead of 4 mA, you could never be sure that it had been adjusted to zero because it could be adjusted too far below zero and you would not be able to see it. So the 4-20 mA signal is a very good choice for process control systems and almost all devices (measurement and control) are available in 4-20 mA.

Differential Pressure Meter   The most common type of flow measurement is a differential pressure meter or transducer (see Figure 13-4). It measures the pressure difference between the inlet and the outlet; therefore, a pressure drop must be created in a pipe. The basic principle of a differential pressure meter is that the pressure drop across the meter is proportional to the square of the flow rate.

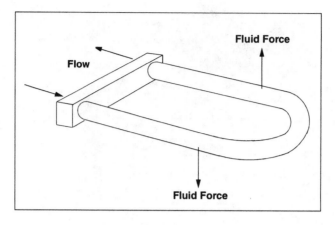

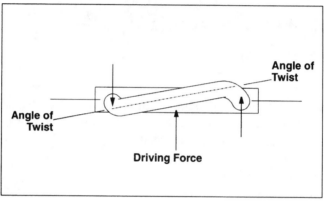

Figure 13–3 Principle of a Coriolis meter. (*Figure published in the Rosemount Product Catalog, Publication No. 00805-0100-1025 © 1998, Rosemount, Inc. Used with permission.*)

Flow Obstruction Devices Numerous devices are used to obstruct the flow and create a pressure drop. An orifice plate is widely used in gas, liquid, and steam flow control applications. It is usually a thin circular plate with a tab by which to handle it. It is normally inserted into a pipe between two flanges, and the pressure drop is measured to determine the fluid's flow rate.

Another device to obstruct flow is a flow nozzle, which is a derivation of an orifice plate but has a larger flow capacity. See Figure 13-5.

A venturi tube also creates a pressure drop (see Figure 13-5). It has a tapered entrance to restrict the flow. The straight section that follows increases the velocity of the fluid, creating a pressure differential between the inlet and outlet that can be measured by a differential pressure meter. A venturi tube can pass 25 to 50 percent more flow than an orifice plate with the same pressure drop. It requires

Figure 13–4 Differential pressure transducer. (*Figure published in the Rosemount Product Catalog, Publication No. 00805-0100-1025 © 1998, Rosemount, Inc. Used with permission.*)

less straight pipe than an orifice plate and is more immune to buildup of scale and corrosion. Because it is very expensive, it is normally used for more demanding or difficult flow control applications.

A *pitot tube* installed in couplings in the pipe creates a differential pressure; it is one of the simplest flow sensors. Bent at a right angle to the flow direction (see Figure 13-5), it measures air flow in pipes and ducts and fluid flow in pipes and open channels. Pitot tubes are simple and reliable and inexpensive, but they are susceptible to clogging.

Positive Displacement Meter A positive displacement meter essentially breaks the fluid into discrete amounts and passes them on (see Figure 13-6). The volume is indicated by the number or rotations of the meter.

A rotary-vane meter is one type of positive displacement meter. It contains an equally divided rotating impeller in the meter's housing that contacts the walls of the housing. The fluid flow causes the impeller to turn. Each segment of the impeller releases a fixed volume of fluid to the outlet. The revolutions of the impeller can then be counted and a flow rate calculated.

Velocity Meters

Turbine Meter A turbine meter has a bladed rotor mounted perpendicular to the flow that causes the blades to rotate. The speed of rotation is directly proportional to the flow rate. The rotational speed can be sensed by a magnetic pickup that counts the pulses to establish the flow rate. A turbine meter is a good choice for accurate liquid measurement applications. They are susceptible to bearing wear.

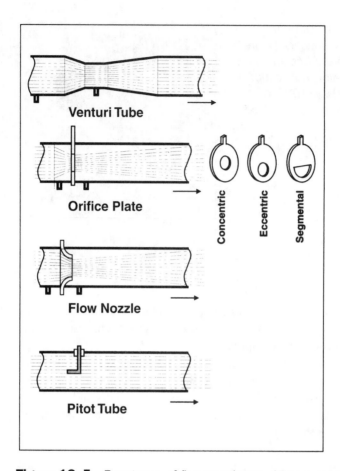

Figure 13–5   Four types of flow restrictors. (*Figure published in the Rosemount Product Catalog, Publication No. 00805-0100-1025 © 1998, Rosemount, Inc. Used with permission.*)

Figure 13–6   Positive displacement meter.

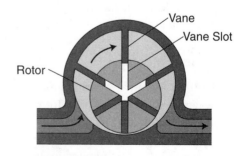

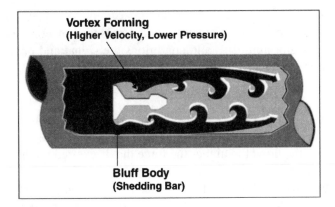

Figure 13–7 Principle of a vortex flowmeter. (*Figure published in the Rosemount Product Catalog, Publication No. 00805-0100-1025 © 1998, Rosemount, Inc. Used with permission.*)

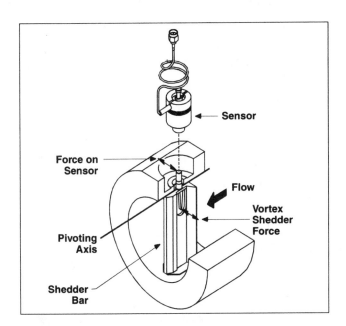

Figure 13–8 Vortex flowmeter. (*Figure published in the Rosemount Product Catalog, Publication No. 00805-0100-1025 © 1998, Rosemount, Inc. Used with permission.*)

Vortex Meter A vortex meter works on the principle of *vortex shedding* (see Figure 13-7). A sensor in the meter senses the presence of vortices and generates pulses. The meter amplifies and conditions the signal and provides an output that is proportional to the flow rate. Figure 13-8 shows a typical vortex flowmeter.

Target Meter A *target meter* measures the force exerted on it by the fluid. Imagine a strain gage attached to a target that is placed in the pipe (see Figure 13-9). The higher the flow, the higher the force that is exerted on the target.

Magnetic Flowmeter A magnetic flowmeter induces an AC field through the flow (see Figure 13-10). The measured AC frequency varies proportionally to the flow rate. This change in frequency is converted to a voltage or current change as an input to the controller. Magnetic flowmeters do not interfere with the flow, nor do they put any restrictions in the line.

An ultrasonic flowmeter with a transmitter and receiver measures flow (see Figure 13-11). The two main types of ultrasonic flowmeters are transit time (pulsed type) and Doppler shift (frequency shift) as shown in Figure 13-11. The transit type utilizes sound waves through the flow. The transducers are mounted

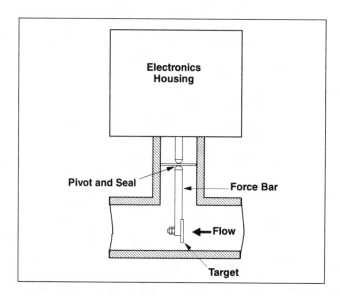

Figure 13–9 Target meter. (*Figure published in the Rosemount Product Catalog, Publication No. 00805-0100-1025 © 1998, Rosemount, Inc. Used with permission.*)

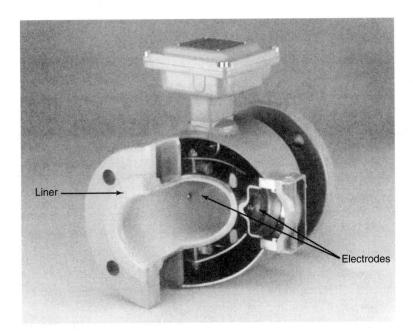

Figure 13–10 Magnetic flowmeters. The left side shows a cutaway section of a flowmeter. The right side shows a typical magnetic flowmeter with a digital transmitter and readout. (*Figure published in the Rosemount Product Catalog, Publication No. 00805-0100-1025 © 1998, Rosemount, Inc. Used with permission.*)

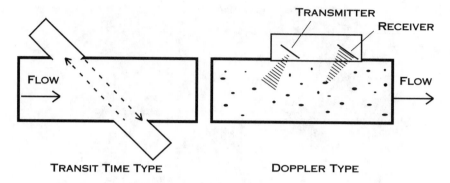

Figure 13–11   Two types of ultrasonic flowmeters.

at either 45 or 90 degrees to the flow. The velocity of the first pulse, directed downstream, increases the speed of the flow. The speed of the next pulse directed upstream is decreased by the flow. The average of the pulse time between the two is proportional to the average velocity of the flow.

Doppler devices measure a shift in frequency caused by the flow of a fluid, which is moving, so it causes a shift in the frequency of the returned pulse. The shift is proportional to the velocity of the fluid. Bubbles, particulates, or any discontinuities cause the pulse to be reflected back to the receiver.

Flow Control System Example

Consider a system that measures differences in pressure to approximate the flow rate. If a restriction is placed in a pipe with flow, the pressure on the downstream side of the restriction will be lower than the pressure on the upstream side of the restriction (see Figure 13-12). An orifice plate can be used to

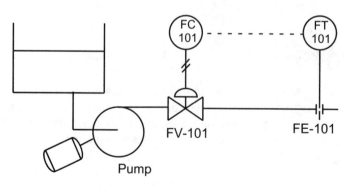

Figure 13–12   Measurement of flow rate by differential pressure measurement.

Figure 13–13   Orifice plate.

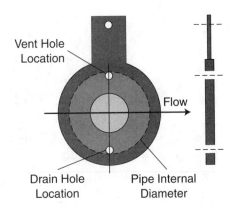

create the pressure differential. If we measure that difference in pressure, we can approximate the flow. The higher the flow, the greater the pressure difference between the two.

The orifice plate (see Figure 13-13) is placed in the pipe. The pressure is then measured on each side of the restrictor plate; it is lower on the downstream side of the plate. One pipe is connected to the upstream side and another is connected to the downstream side. These two pipes are then connected to a device called a *differential pressure transducer (DPT)*. The DPT is shown as FT-101 (flow transmitter 101). A DPT has two inputs, a high pressure side and a low pressure side. Imagine a membrane in the center of these two pressure ports. If there is a pressure difference between the two, the membrane yields to the low pressure side. If the membrane has a strain gage, it can measure the difference in pressure. The DP then converts the strain gage signal into a 4-20 mA signal. The dotted line in the system drawing shows that the signal from FT-101 is connected to FC101 (flow controller 101). The 4-20 mA becomes the feedback signal to a controller. This has a problem, however. This measures pressure with the DPT but we really want to measure flow. This would not be a problem if a linear relationship exists between pressure and flow, but it does not. The pressure increase is the square of the flow increase. Fortunately, we can buy DPTs or signal conditioners *square root extractors* that take the pressure reading and the square root of the difference and converts the result to a 4-20 mA signal.

Next the flow controller (FC-101) must control the flow rate. We can use a proportional valve to control the flow rate (see Figure 13-14). We have now added a proportional valve to the flow control system. The proportional valve opens between 0 and 100 percent, depending on the input it receives. The valve in this system requires a 4-20 mA input signal. At 4 mA the valve is closed; at 20 it is 100 percent open. The 4-20 mA signal is converted to air pressure to move a shaft and control the position of the valve.

Figure 13–14 Proportional valve.

Flow Control of Solids

Many industrial processes need to control the flow of solids (particulates), especially when batching for recipe control. Solid flow control usually involves weight measurement. Figure 13-15 is a simple example of a flow control system for solids. Flow in weight systems is expressed as weight per unit of time:

$$\text{Flow rate} = \text{Weight} \times \text{Time}$$

Assume that the particulates are relatively consistent and that a uniform amount flows from the hopper onto the conveyor. The controller is able to control the flow by adjusting the drive speed. A faster rate of speed on the conveyor allows more particulates to drop from the hopper onto the conveyor. In this case, the weight per unit of time increases. The weight of the particulates is measured on the conveyor belt. The weight measurement must be taken a sufficient distance from the hopper so that the material is not in a turbulent state from dropping out of the hopper and onto the conveyor, but has settled on the conveyor. The weight station must be po-

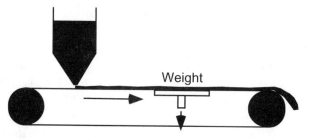

Figure 13–15 Simple flow control system for solids.

sitioned so that it will not be influenced by a lifting effect. In other words, conveyors have areas where the belt may actually be lifted partially off the weight station. This would cause erratic and inaccurate feedback and thus output.

The weigh stations utilize load cells (strain gage) or linear variable displacement transducer (LVDTs). The load cells are mounted under a weigh station and generate an output base that is proportional to the weight of the particulate on the weigh station. LVDTs can measure the droop in the conveyor. The heavier the load, the more the conveyor droops. The more the conveyor droops, the more the LVDT is displaced. The distance that the LVDT is displaced is proportional to the weight of the particulates on the conveyor.

LEVEL CONTROL

Level control for liquids or solids is very common in industry. *Level* is the interface between two material phases. The different phases that need to be measured and controlled could be between two solids, a liquid and a solid, a liquid and a gas, two liquids, and so on. Level is controlled using a reference level (setting). Many level control applications are very simple and require only rudimentary control. For example, a switch that senses only whether the level is above or below the reference level. Many other applications, however, require closer control.

Level Control Methods

Bubbler Method The bubbler method is one common method to measure a liquid level (see Figure 13-16). This method uses one or two tubes. The basic principle is very simple. Imagine blowing air through a tube into a tank of water. The deeper the tube is placed in the water, the more pressure it takes to blow bubbles. The higher the tube is raised, the easier it is to blow bubbles.

The bubble method puts air pressure into the bubbler tube. Just enough pressure is applied to produce one or two bubbles per second when the fluid is at the desired level. A simple needle valve controls the air pressure or a constant airflow device more closely controls the pressure that is allowed into the tube. The pressure in the bubbler tube can then be measured by a pressure switch or a pressure transmitter. Note that the higher the level of fluid, the higher the pressure of the fluid in the bottom of the tank. This means that fewer bubbles escape and the pressure in the tube is higher. A low level of water decreases the pressure of the fluid at the bottom of the tank, and bubbles escape more readily, decreasing the pressure in the tube. The pressure transmitter can measure these changes in bubbler tube pressure and transmit the values to a controller.

Many applications require two tubes, as in enclosed, pressurized tanks. One tube is located at the bottom of the tank and the other at the top of the tank (see

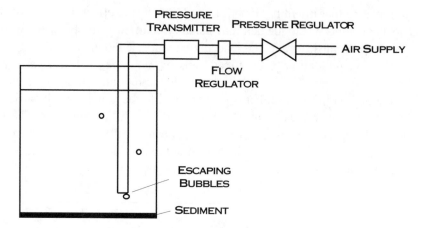

Figure 13–16 Bubbler method of level measurement. It may be necessary to keep the bottom of the bubbler tube above any sediment that is present in the tank.

Figure 13-17). A differential pressure transmitter, discussed earlier in the flow control section is used in this case. The lower tube pressure is attached to the high pressure side, and the tube at the top of the tank is connected to the low pressure side. The difference in these pressures can be used to measure density. It is usually necessary to provide a cleanout in the bubbler tubes. A disadvantage of this

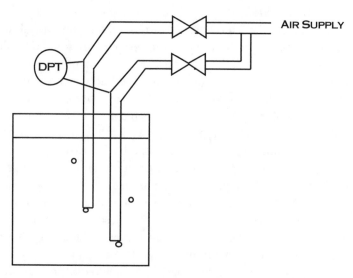

Figure 13–17 Bubbler method of density measurement.

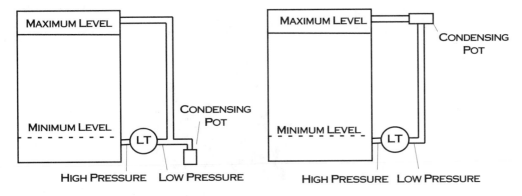

Figure 13–18 A dry-leg system on the left and a wet-leg application on the right.

pipe is the buildup of solids in the bubbler tubes. Buildup in the tube negatively impacts the system's accuracy.

See Figure 13-18 for two types of level applications. The application on the left is called a *dry leg*. Its top pressure tube is exposed only to air in the system. Condensation and collection in the low pressure side distorts the reading. A condensing pot below the low pressure input side to the differential pressure transducer collects any condensate, which must be drained periodically. The example on the right side of Figure 13-18 shows a wet-leg application. This application has a vapor over the fluid in the tank. The vapor in the low pressure side distorts the reading. In this case a condensing pot is placed at the top of the dry leg to condense and remove the vapor that would distort the reading.

The bubbler method requires the fluid to be of somewhat constant density. The temperature or the mix of the fluids in the tank can affect the density. To illustrate this, imagine a tank of molasses. If the temperature is 150 degrees Fahrenheit, the bubbles flow quite freely. If the temperature is 40 degrees Fahrenheit, there are few bubbles. In this case the temperature should be kept somewhat constant.

Ultrasonic Sensor Method Ultrasonic sensors also can be used to measure the level of liquids or solids. A sound wave is transmitted into the tank and reflected off of the solid or liquid back to the sensor. The amount of time it takes to return is converted to an analog signal that represents the level of the tank. There are some problems with the use of ultrasonics. Air temperature affects the performance of ultrasonics, but this can be compensated for. Ultrasonics are not appropriate for fluids foaming or bubbling, as in a boiling application because they can produce inaccurate readings. They may also be inappropriate for solids whose levels are not consistent (see Figure 13-19).

Figure 13–19 Level of solids used in this tank is not consistent.

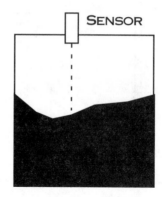

Radar Method   A radar gage can be used to provide level feedback (see Figure 13-20). Notice that this gauge can be used in difficult applications that have high agitation and/or foaming.

Weight Method   Load sensors can be used to measure the weight of a tank. Obviously, the more material a tank contains, the heavier it is. A weight measurement level system is illustrated in Figure 13-21. A load cell transmitter (Figure 13-22) placed under each column supports the tank and sends a signal to the controller. The heavier the load, the larger the signal. The sensor use is based on the principle of the strain gage. The weight method of level measurement is usually more expensive and more difficult to design and build than other level measuring

Figure 13–20 Radar gauge measuring the level of the fluid in the tank. (*Figure published in the Rosemount Product Catalog, Publication No. 00805-0100-1025 © 1998, Rosemount, Inc. Used with permission.*)

Figure 13–21 Weight
method system.

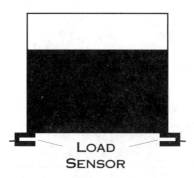

LOAD
SENSOR

devices. The input and output piping cannot interfere with the upward and downward movement of the tank.

Capacitive Level Sensing Method Capacitive sensors can also measure a liquid or solid level. See Figure 13-23 for examples of capacitive-level measurement for conductive and nonconductive materials. The insulated metal probe is one side of the capacitor and the tank is the second side. The capacitance varies as the level around the probe varies. A capacitance bridge measures the change in capacitance.

Radioactive Method Figure 13-24 shows two examples of radioactive level measurement. Radioactive methods use cobalt 60 or cesium 137 to provide a source of radiation. The radiation is emitted on one side of the tank and sensed on the other. The higher the level, the more radiation is absorbed before it reaches the sensor. The amount of radiation sensed can be converted to a level reading. The level of radiation is low, and the danger is relatively low, but the use of radiation technology requires its operator to be a licensed technician.

Figure 13–22 Load
sensor.

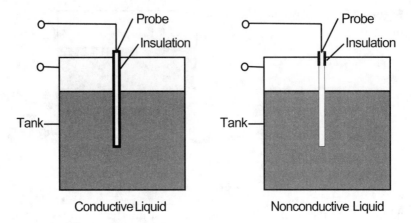

Conductive Liquid   Nonconductive Liquid

Figure 13–23   Capacitive level sensor.

Level Control Example

Density Control   To explain density control, we use chocolate milk as our product. The process is quite simple: mix milk with chocolate syrup. The densities of milk and chocolate syrups are quite different. This is a good example of a density control system. Accomplishing this task requires measuring and controlling the density of the product. First, we would mix batches until we find the perfect taste. Then we could measure the density of the perfect batch and use that as our setpoint.

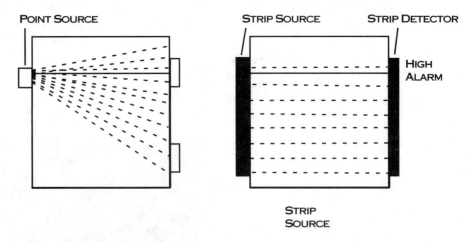

Figure 13–24   Radioactive level measurement. Point and strip source radiation level measurement.

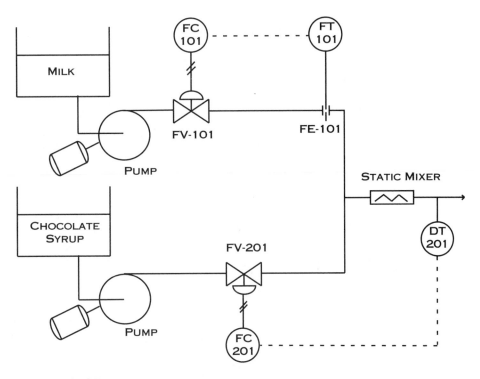

Figure 13–25 Density control system.

This system uses a Coriolis meter to measure the product's density because it can measure extremely small changes in density. We will control the flow of milk and syrup into the process (see Figure 13-25) using a static mixer to mix our milk and syrup thoroughly (see Figure 13-26). The Coriolis meter measures the density after using the static mixer.

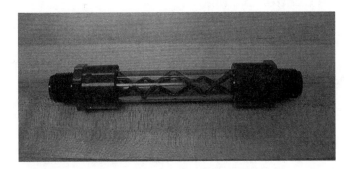

Figure 13–26 Static mixer.

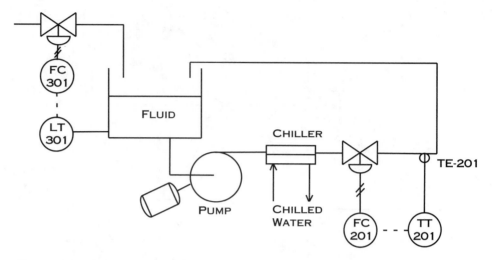

Figure 13–27   Temperature control system

The first step is choosing a desired flow rate for the milk, assumed here to be 1 gallon per minute. The process controller controls a valve to set a flow of 1 gallon per minute for the milk.

The next step is to control the flow rate of the chocolate syrup accurately. The process loop (density) controls the flow of syrup into the process to keep the density of the product at setpoint. The Coriolis meter provides feedback on the density to the controller. The controller controls the valve position for the chocolate syrup to ensure the correct density of chocolate milk.

TEMPERATURE CONTROL

Temperature Control Example

This process controls the temperature of a fluid in a tank. Heated fluid comes into the tank and then must be cooled to 95 degrees for the specific process. Actually, this process has two loops: a level control loop and a cooling loop (see Figure 13-27). Note that each loop has a specific range of numbers. The chiller loop, for example, is 201. The level loop is 301.

Figure 13–28   ISA connection symbols.

Figure 13–29   Heat exchanger. Two ports are used for hot water from the tank and two are used for the chilled water.

The *level control loop* consists of an inductive transducer and transmitter to measure the height of fluid. The level sensor's signal is sent to the controller, which controls a ball valve that regulates the flow of fluid into the tank.

The level control loop also consists of a level transmitter (LT-301) and a flow controller (FC-301) that regulates a valve. The dashed lines connecting the flow controller to the valve indicate an air line. In this system heated fluid is allowed into the tank until reaching the desired level. Then the cooling process begins. Figure 13-28 shows a few ISA connection symbols.

The *cooling loop* consists of a pump, a ball valve, an RTD and a heat exchanger. The pump circulates the water from the tank through the heat exchanger loop, and a ball valve controls the flow through it. The RTD temperature element 201 (TE-101) and the temperature transmitter (TT-201) provide temperature feedback. The output from the temperature transmitter is connected to the flow controller (FC-201) to control the ball valve (with a pneumatic signal). See Figure 13-29. Chilled water and heated tank water are circulated through the heat exchanger separately. The cooled water returns to the tank. This continues until the temperature reaches the setpoint in the tank when the tank is emptied so a new batch can be processed.

CALIBRATION

The accuracy and efficiency of any process control system depends on each device in the loop. Calibration is performed to ensure that the output from each device accurately represents the input.

In a manufacturing plant all measurement devices that affect quality are regularly calibrated because they can be dropped and damaged, or they may simply become inaccurate over time. Calibration should be performed when a device is first installed, after a repair, and on a regular basis. A sticker is placed on each measurement and control device telling when the next calibration is required, and the device cannot be used after that date until it has been calibrated.

Calibration Example

A device such as a micrometer would be taken to the quality department where its accuracy would be checked against a very accurate micrometer standard. If it is inaccurate, it is adjusted; if it cannot be made accurate, it is disposed of. The standard also must be checked on a regular basis against an even more accurate standard to ensure its accuracy.

Measurement and control devices usually receive an analog input and transmit an analog output based on the input value. For example, a typical pressure-to-current converter takes an input pressure between 3 and 15 pounds of pressure. The span (or range) of input values is 12 pounds (15 – 3). It outputs 4-20 milliamps over the 3–15 pound input pressure. The output span (or range) is 16 milliamps.

A calibration test is performed from the bottom to the top input value (Figure 13-30) and then from the top to the bottom value (Figure 13-31). This ensures that the device is accurate in both directions and shows any effects that may be due to hysteresis. Values of 0 percent or 100 percent should not be used because they cannot be approached from below 0 percent or above 100 percent.

For the upscale calibration procedure, a technician begins at zero pressure and proceeds up the scale input test values. This helps eliminate the possible effects of

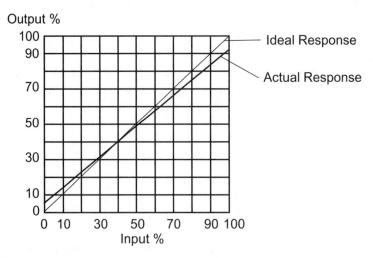

Figure 13–30   Values for upscale procedure.

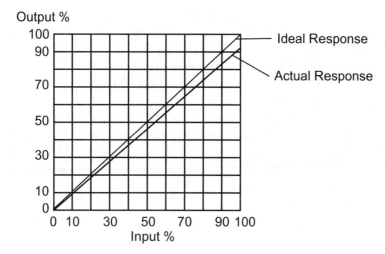

Figure 13–31 Values for downscale procedure.

hysteresis. If the technician adjusts the input too high for an upscale measurement, the value must be reduced and approached from below the test value to ensure accurate calibration. The downscale procedure is performed in the opposite manner: The technician starts with 100 percent pressure and reduces pressure down to each required input value. If the technician adjusts the input too low for a downscale measurement, the value must be increased and approached from above to ensure accurate calibration.

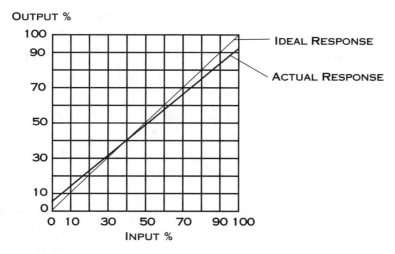

Figure 13–32 Graph of actual output vs. ideal output.

Usually two adjustments are made to devices: the zero and the span adjustments. To correctly calibrate this instrument, the technician first gives the device a 0 input and adjusts it to the 10 percent value and then adjusts the zero adjustment until its output is correct. The technician next increases the input value to 90 percent and then adjusts the span adjustment until the output value is correct.

The technician then must recheck the 10 percent value to ensure that it has not changed. If it has changed, the technician must perform both adjustments again. Next a simple graph is made of the input versus output values (Figure 13-32).

The graph in Figure 13-32 indicates two errors: zero and span. The measured value for the actual output response is not at 0,0 on the graph but is shifted up. The zero needs to be adjusted. The actual response at 100 percent input is too low, meaning the span is too small.

QUESTIONS

1. List the three main types of flow.
2. Explain the principle of the measurement of flow rate by measuring differences in pressure.
3. What is an orifice plate?
4. What is a flow nozzle?
5. What is a venturi?
6. Name one type of mass flowmeter and explain the principle of its operation.
7. Describe the purpose and use of a differential pressure transducer.
8. Describe the bubbler method of level measurement.
9. Explain the principle of density measurement and control.
10. Explain the similarities between different types of process control systems.
11. When performing an upscale test, the input test point should be approached from
    a. above
    b. below
    c. either direction
12. If zero shift and a span error are detected, the first to be corrected is usually
    a. zero
    b. span

c. either
d. neither

13. The following graph shows the results of a calibration procedure. Explain the sequence of adjustments to be made.

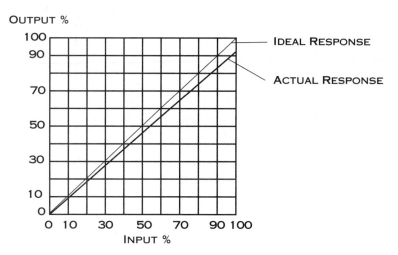

14. The following graph shows the results of a calibration procedure. Explain the sequence of adjustments to be made.

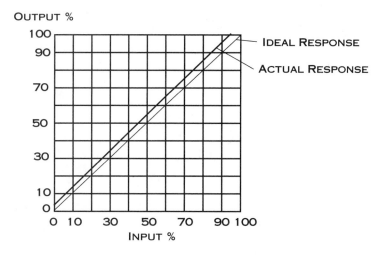

15. The following graph shows the results of a calibration procedure. Explain the sequence of adjustments to be made.

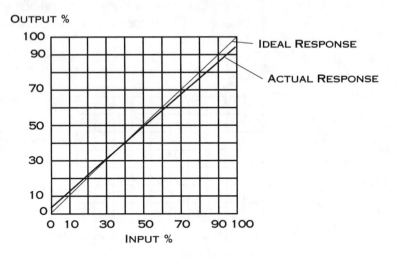

chapter
Overview of Plant Floor Communication
14

If enterprises are to become more productive, they need to improve processes. This requires accurate data. Communications are vital. Production devices hold very valuable data about their processes. In this chapter we examine how these data can be acquired.

OBJECTIVES

Upon completion of this chapter you will be able to:
1. Define the terms *serial communication, RS-232, RS-422,* and *SCADA software*.
2. Describe how computers can communicate with PLCs.
3. Describe several industrial networks.

SHOP FLOOR COMMUNICATION

The programmable logic controller (PLC) has revolutionized manufacturing. It has made automation flexible and affordable. It controls processes across the plant floor. In addition to producing product, it also produces data that can be more profitable than the product. Most processes are very inefficient. If we can use the data to improve processes, we can drastically improve profitability. These inefficiencies are not normally addressed because people are busy with other, seemingly more pressing problems. (A good friend of mine in a small manufacturing facility said it best: "It's hard to think about fire prevention when you're in the middle of a forest fire.") Manufacturing people are usually amazed to find that most of the data they would like to have about processes is already being produced in the PLCs or other smart devices used in their processes. With very few changes, the data can be gathered and used to improve quality, productivity, and uptime. Huge gains are possible if these data are used.

First the data must be acquired. Many managers today will say that they are already collecting much of the data from the plant floor manually. But data gathered manually are often very inaccurate and are not real time. The information must be written down by an operator, gathered by a supervisor, taken "upstairs," entered by a data processing person, printed into a report, distributed, and analyzed.

This all takes time, making the data often many days or weeks late. If mistakes in entry were made on the floor, they are seldom corrected. The lateness and inaccuracy of the data gathering makes it almost counterproductive.

Real-time information is required about the accuracy of the operator's orders, instructions, current specifications, and so on. This information is often lacking in industrial and service enterprises today. The use of electronic communications quite easily achieved this communication. Many of the data required already exist in the smart devices on the factory floor. Much of the data that people write on forms in daily production already exist in PLCs.

Improvements in computer hardware and software have made communication much easier. Many software communications packages make it easy to communicate with PLCs.

Primitive Communications

Some devices such as simple PLCs do not have the capability to communicate. For example, they cannot communicate serially with other devices. This results in the use of primitive methods restricting the devices to essentially just a handshake with a few digital inputs and outputs. For example, a robot is programmed to wait until input 7 comes true before executing program number 13. It is also programmed to turn on output 1 after it completes the program. A PLC output can then be connected to input 1 of the robot, and output 1 from the robot can be connected to an input of the PLC. This is a simple one-device cell with primitive communications. The PLC can command the robot to execute. When the program is complete, the robot notifies the PLC with very simple yes or no (binary) information.

Serial Communications

Communications capability, such as the need to upload/download programs or update variables cannot be done with primitive communications. Most machines offered serial communications capability using the asynchronous communications mode with an RS-232 serial port available. An RS-232 serial port does not

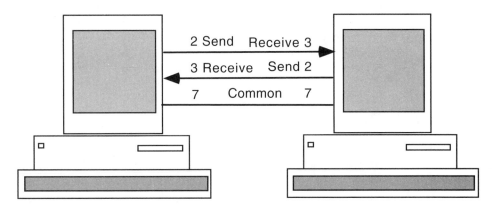

Figure 14–1 Simple RS-232 wiring with the simplest of RS-232 connections.

easily communicate with any other device with an RS-232 port, however. Each machine could have its own protocol.

The RS-232 standard specifies a function for each of 25 pins although some manufacturers use only three, as illustrated in Figure 14-1. Some manufacturers use more than three pins, so some electrical handshaking can take place. In Figure 14-1 no handshaking occurs. The first computer sends a message whether or not there is another machine. The computer cable could be unplugged or the computer turned off; the sender does not know. Handshaking implies a cooperative operation. The first computer tells the second that it has a message it would like to transmit by setting pin 4 (the request to send pin) high. The second machine sees the request to send pin high, and if it is ready to receive, it sets the clear to send pin high. The first computer then knows that the cable is connected, the computer is on, and it is ready to receive. Some devices can be set up to handshake; others cannot.

Fortunately when a machine is purchased, it is generally capable of communicating with an IBM personal computer. The user usually uses the section on communication in the device manual to find a pinout for the proper cable. Communication when a wide variety of devices is involved can be difficult and expensive.

A message sent using asynchronous communications is broken into individual characters and transmitted one bit at a time normally using the ASCII system. Every ASCII letter, number, and some special characters have a binary-coded equivalent. A 7-bit ASCII has 128 possible different letters, numbers, and special characters. An 8-bit ASCII has 256.

Each character is sent as its ASCII equivalent (for an example, see Figure 14-2). It takes 7 more bits to send a character in the asynchronous model, however. Other bits are used to make sure the receiving device knows that a message is coming, that the message was not corrupted during transmission, and that the

Figure 14–2 Letter A transmitted in the asynchronous serial mode of communications with odd parity. The character A has an even number of ones, so the parity bit is a 1 to make the total odd. If the character to be sent has an odd number of ones, the parity bit is a 0. The receiving device counts the number of 1s in the character and checks the parity bit. If they agree, the receiver assumes that the message was received accurately. This is rather crude error checking because two or more bits could change state; the parity bit could still be correct but the message wrong.

character has been sent. The first bit sent is the start bit (see Figure 14-3) to let the receiver know that a message is coming. The next 7 bits (8 if 8-bit is used) are the ASCII equivalent of the character. Then a bit is reserved for parity, which is used for error checking. The parity of most devices can be set up for odd or even, mark or space, or none.

Some new standards help integrate devices more easily. RS-422 and RS-423 were developed to overcome some of the weaknesses of RS-232. Their distance and speed of communications are drastically higher.

RS-422 is called *balanced serial*. RS-232 has only one common wire. The transmit and receive lines use the same common wire. This can lead to noise problems. RS-422 solves this problem by having separate commons for the transmit and receive lines, making each line noise immune. The balanced mode of communications exhibits lower crosstalk between signals and is less susceptible to external interference. Crosstalk is the bleeding of one signal onto another. This reduces the potential speed and distance of communications. This is one reason that the distance and speed for RS-422 is much higher. RS-422 can be used at

Start Bit	Data Bits	Parity Bits	Stop Bit
1 Bit	7 or 8	1 Bit (Odd, Even, Mark, Space, or None)	1, 1.5, or 2 Bits

Figure 14–3 Transmission of a typical ASCII character.

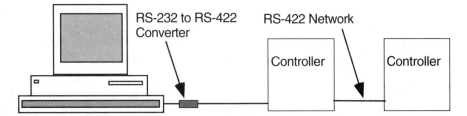

Figure 14–4 Use of a converter to change RS-232 communications to RS-422 communications. The computer then is able to communicate with other devices on its network. The devices on the right are on an RS-422 network.

speeds of 10 megabits for distances of over 4000 feet, compared to 9600 baud and 50 feet for RS-232.

RS-423 is similar to RS-422 except that it is unbalanced, and has only one common wire that the transmit and receive lines must share. RS-423 allows cable lengths exceeding 4000 feet. It is capable of speeds up to about 100,000 bauds per second.

RS-449 is the standard that was developed to specify the mechanical and electrical characteristics of the RS-422 and RS-423 specifications. The standard addressed some of the weaknesses of the RS-232 specification. The RS-449 specification specifies a 37-pin connector for the main and a 9-pin connector for the secondary. Remember that the RS-232 specification does not specify what type of connectors or how many pins must be used.

These standards will eventually cause RS-232 to be replaced, but there are so many of them that this will take a long time, although it is already occurring rapidly in industrial devices. Many PLCs' standard communications are performed with an RS-422 standard.

Adapters are cheap and readily available to convert RS-232 to RS-422 or vice versa (see Figure 14-4). This conversion can be used to an advantage if a long cable length is needed for an RS-232 device (see Figure 14-5).

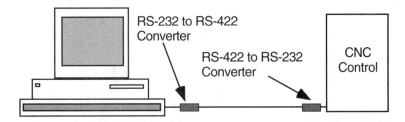

Figure 14–5 Two converters used to extend the cable length. Remember that RS-232 is only reliable to about 50 feet. The use of two converters allows 4000 feet to be covered by the RS-422 and then converted to RS-232 on each end. The speed will be limited by the RS-232.

RS-485 is a derivation of the RS-422 standard. The main difference is that it is a multidrop protocol that can have many devices on the same line. This requires that the devices have some intelligence, however, because they must each have a name so that each knows when it is being talked to. Many PLCs and other smart devices now utilize RS-485 protocol. The standard specifies the electrical characteristics of receivers and transmitters connected to the network. The standard specifies a differential signal between −7 to +12 volts. The standard limits the number of stations to 32. This allows for up to 32 stations with transmission and reception capability, or 1 transmitter and up to 31 receiving stations.

This communication is usually accomplished by the use of supervisory control and data acquisition (SCADA) software. The concept is that software is run in a common microcomputer to enable communications to a wide variety of devices. The software is typically similar to a generic building block (see Figure 14-6).

The programmer writes the control application from menus or in some cases graphic icons and then loads drivers for the specific devices in the application. A driver is a specific software package written to handle the communications with a specific brand and type of device. It is available for most common devices and are relatively inexpensive.

The main task of the software is to communicate easily with a wide range of brands of devices. Most enterprises do not have the expertise required to write software drivers to communicate with devices, but they can use SCADA packages to simplify this task. In addition to handling the communications, SCADA software makes it possible for applications people who are not programmers to

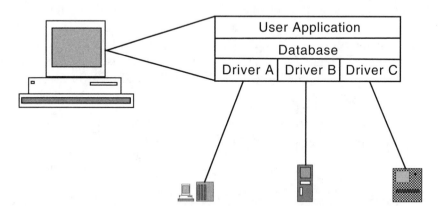

Figure 14–6 The user application of a typical SCADA software package defines which variables from the devices must be communicated. These are collected through the drivers and stored in a database that is available to the application. Once the computer has the desired data, it is a relatively easy task to make the data available to other devices.

write the control programs. This allows the people who best know the application to write it without learning complex programming languages. Drivers are available for all major brands of PLCs and other common manufacturing devices.

In general, an applications person would write the specific application using menu-driven software which is easy to use. Some are like spreadsheets and some use icons for programming. The programmer uses tagnames instead of specific I/O numbers that the PLC uses. For example, the application might involve temperature control. The actual temperature might be stored in register S20 in the PLC. The applications person would use a tagname such as Temp_1 instead of the actual number (see Figure 14-7).

This makes the programming *transparent*, meaning that the application programmer does not have to worry much about what brands of devices are in the application. A table is set up to assign specific PLC addresses to the tagnames (see Figure 14-7). In theory, if a different brand PLC were installed in the application, the only change required would be a change to the tagname table and the driver. Fortunately, more software is available daily to make the task of communications easier. The software is faster, more friendly, more flexible, and more graphics oriented. The operator can use the data gathered by SCADA packages for statistical analysis, historical data collection, adjustment of the process, or graphical interface.

Local Area Networks (LANs)

Local area networks (LANs) are the backbone of communications networks. LANs can be broken down into three various methods of classification for LANs: topology, cable, and access.

Topology *Topology* refers to the physical layout of LANs. The three main types are star, bus, and ring.

Device	Actual Number	Tagname
PLC 12	REG20	Temp_1
PLC 12	REG12	Cycletime_1
PLC 10	S19	Temp_2
PLC 07	N7:0	Quantity_1
Robot 1	R100	Quantity_2

Figure 14–7 What a tagname table might look like.

Figure 14–8   Star topology.

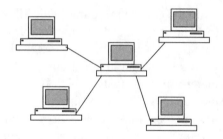

Star Topology   The star style uses a hub to control all communications. All nodes are connected directly to the hub node (see Figure 14-8). All transmissions must be sent to the hub, which then sends them on to the correct node. One problem with the star topology is that if the hub goes down, the entire LAN is down.

Bus Topology   The bus topology is a length of wire into which nodes can be tapped. At one end of the wire is the head end (see Figure 14-9), which is an electronic box that performs several functions. In a broadband bus the head end receives all communications and then remodulates the signal and sends it out to all nodes on another frequency. *Remodulate* means that the head end changes the received signal frequency to another and sends it out for all nodes to hear. Only the nodes that are addressed pay attention to the message. The other end of the wire (bus) dissipates the signal. Many industrial networks are based on bus technology.

Ring Topology   The ring topology looks like it sounds. It has the appearance of a circle (see Figure 14-10). The output line (transmit line) from one computer goes to the input line (receive line) of the next computer, and so on. It is a very straightforward topology. A node sends a message out on the transmit line. The

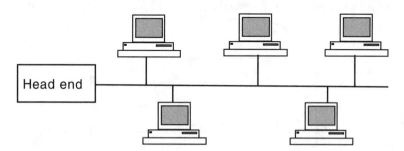

Figure 14–9   Typical bus topology. Each node (communication device) can speak on the bus. The message travels to the head end and is converted to a different frequency. It is then sent back, and every device receives the message. Only the device that the message was intended for pays attention to the message.

Figure 14–10   Ring topology.

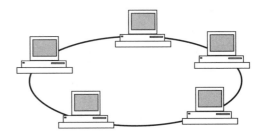

message travels to the next node. If the message is addressed to it, the node writes it down; if not, it passes it on until the correct node receives it.

The more likely configuration of a ring is shown in Figure 14-11. This style is still a ring; it just does not appear to be. This is the convenient way to wire a ring topology. The main ring (backbone) is run around the facility, and interface boxes are placed in line at convenient places around the building. These boxes are often placed in "wiring closets" near where a group of computers will be attached. These interface boxes, called *multiple station access units* (MSAUs), are just like electrical outlets. A computer can be plugged into an outlet on the MSAU. The big advantage of the MSAU is that devices can be attached/detached without disrupting the ring or communications.

Cable Types   The four main types of transmission media are twisted pair, coaxial, fiber-optic, and radio frequency. Each has distinct advantages and disadvantages. The capabilities of the cable types are expanding continuously.

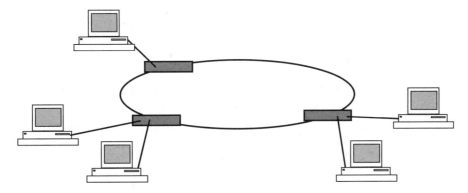

Figure 14–11   Typical ring topology. Note that it does not look like a ring. It actually is a ring, but it looks more like a star topology. The multiple station access units (MSAUs) allow multiple nodes to be connected to the ring. The MSAU looks like a set of electrical outlets into which the computers are plugged and then are attached to the ring. The MSAU has relays for each port. This allows devices to be plugged in and removed without disrupting the ring.

Twisted Pair Twisted-pair wiring is, as its name implies, pairs of conductors (wires) twisted around each other along their entire length. The twisting of the wires helps make them more noise immune. The telephone wires that enterprises have throughout their buildings are twisted pair. There are two types of twisted-pair wiring: shielded and unshielded. The shielded type has a shield around the outside of the twisted pair to help make the wiring noise immune. The wiring that is used for telephone wiring is typically unshielded. The newer types of unshielded cable are more noise immune than in the past.

Coaxial Cable Coaxial or coax cable is a very common communication medium. Cable TV uses coaxial cable. It is broadband, which means that many channels can be transmitted simultaneously. It has excellent noise immunity because it is shielded (see Figure 14-12).

Broadband technology is more complex than *baseband* (single channel). Broadband technology has two ends to the wire. One end is called the *head end*. It receives all signals from devices that use the line. The head end then remodulates (changes to a different frequency) the signal and sends it back out on the line. All devices hear the transmission but pay attention only if it is intended for their address.

Broadband technology uses *frequency-division multiplexing*. The transmission medium is divided into channels, and each channel has its own unique frequency. There are also buffer frequencies between each channel to help with noise immunity. Some channels are for transmission and some are for reception.

Time-division multiplexing, also called *time slicing*, is used in the baseband transmission method. Several devices may wish to talk on the line (see Figure 14-13) but cannot wait for one device to finish its transmission completely before another begins. They must share the line. One device takes a slice of time, then the next does, and so on.

Fiber-Optic Cable Fiber-optic technology is also changing very rapidly (see Figure 14-14 for the appearance of the cable.) The major disadvantage of fiber-optics are its complexity of installation and high cost; however, the installation has become much easier and the cost has fallen dramatically to the point that fiber's total cost is not much different than some installations of shielded twisted pair.

The advantages of fiber are its perfect noise immunity, high security, low attenuation, and high data transmission rates. Fiber transmits with light, so it is un-

Figure 14–12 Coaxial cable with shielding around the conductor.

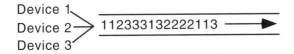

Figure 14–13 Each device in a time-division multiplexing must share time on the line. Device 1 sends part of its message and then gives up the line so that another device can send, and so on.

affected by electrical noise. The security is good because fiber does not create electrical fields that can be tapped like twisted pair or coaxial cable. The fiber must be physically cut to steal the signal, making it a much more secure system.

All transmission media attenuate signals, meaning that the signal gets progressively weaker the farther it travels. Fiber exhibits far less attenuation than other media. It can also handle far higher data transmission speeds than can other media. The fiber distributed data interchange (FDDI) standard was developed for fiber cable. It calls for speeds of 100 megabits, which seemed very fast for a short period of time, but many believe that it may be possible to get 100 megabits with twisted pair, so the speed standard for fiber may be raised.

Radio Frequency Radio frequency (RF) transmission has recently become very popular. Its use has exploded in the factory environment. The major makers of PLCs have RF modules available for their products. These modules use radio waves to transmit the data. They are very noise immune and perform well in industrial environments. RF is especially attractive because no wiring, which is expensive to install, needs to be run and wires are susceptible to picking up electrical noise. RF is less difficult and less expensive than wire to move devices once they are installed. Wireless local area networks (WLANs) are finding a home everywhere from the factory floor to the office.

Wireless technology is becoming very transparent; it has become virtually a "black box." A wireless network operates exactly the same as a hardwired network. Devices are simply attached to a transceiver (combination transmitter/receiver). The transceiver performs all of the translation and communication necessary to convert the electrical signals from the device to radio signals and then sends them via radio waves. The radio waves are received by another transceiver, which translates them back to network signals and puts them back on the network.

A wireless LAN has two parts: an access transceiver and remote client transceivers. The access transceiver is typically a stationary transceiver that attaches to

Figure 14–14 Fiber-optic cable with multiple fibers through one cable.

the main hardwired LAN. The remote client transceivers link the remote parts of a LAN to the main LAN via radio waves. These transceivers can be used as *bridges* or *gateways*. A bridge links two networks that utilize the same or very similar protocols. A gateway is more like a translator; it connects dissimilar network protocols so that two different kinds of networks can communicate.

Radio transmissions use spread spectrum technology, which was developed by the U.S. military during World War II to prevent radio signal jamming and *also* to make them difficult to intercept. Spread spectrum technology uses a wide frequency range. It is very noise immune.

The transmitted signal bandwidth in spread spectrum technology is much wider than the information bandwidth. Radio stations utilize a narrow bandwidth for transmission. The transmitted signals (music and voice) utilize virtually all of the bandwidth. In spread spectrum the data being translated is modulated across the wider bandwidth. The transmission looks and sounds like noise to unauthorized receivers.

A special pattern or code determines the actual transmitted bandwidth. Authorized receivers use the codes to pick out the data from the signal.

The FCC has dedicated three frequency bands for commercial use: 900 MHz, 2.4 GHz, and 5.7 GHz. These frequencies have very little industrial electrical noise.

Access Methods

Token Passing In the token-passing method, only one device can talk at a time. The device must have the token to be able to use the line. The token circulates among the devices until one of them wants to use the line (see Figure 14-15). The device then grabs the token and uses the line.

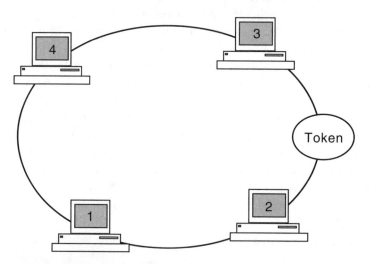

Figure 14–15 Token-passing method.

The device that wants to talk waits for a free token. The sending station sets the token busy bit, adds an information field, adds the message it would like to send, and adds a trailer packet. The header packet contains the address of the station for which the message was intended. The entire message is then sent out on the line. Every station examines the header and checks the address to see if it is being talked to. If not, it ignores the message. The message arrives at the receiving station and is copied. The receiving station sets bits in the trailer field to indicate that the message was received. It then regenerates the message and sends it back out on the line. The original station receives the message back and sees that the message was received. It then frees the token and sends it out for other stations to use.

Token passing offers very reliable performance and predictable access times, which can be very important in manufacturing. Predictable access is often called *deterministic* because actual access times can be calculated based on the actual bus and nodes.

OVERVIEW OF INDUSTRIAL NETWORKS

The capabilities of field devices have increased rapidly as has their ability to communicate. This has led to the need for them to network. Imagine what it would be like in your home if there was no electrical standard. Your television might use 100 volts at 50 cycles. Your refrigerator might use 220 at 50 cycles, and so on. Every device might use different plugs. It would be a nightmare to change any device to another brand because the wiring and outlets would need to be changed. Industrial networks are an attempt to create standards to allow different brands of field devices to communicate and be used interchangeably.

The hierarchy of typical communications in an enterprise can be divided into three levels: information or enterprise, automation and control, and device. Rockwell Automation has a three-level model of communications (see Figure 14-16). The information or enterprise level is for plantwide data collection and maintenance. The second level is for automation and control. It is used for real-time I/O control, messaging, and interlocking. The lowest level is the device layer, which is used to cost effectively integrate low-level devices into the overall enterprise. Figure 14-16 shows an Ethernet network at the information (enterprise) level. ControlNet™ is used at the second level for automation and control. The device level utilizes DeviceNet to communicate.

This chapter primarily focuses on the device level, although it briefly examines the control level and enterprise level. The discussion begins at the lowest (device level) and moves up in the communications hierarchy. It is very important to note that the three levels are not always present or clearly defined.

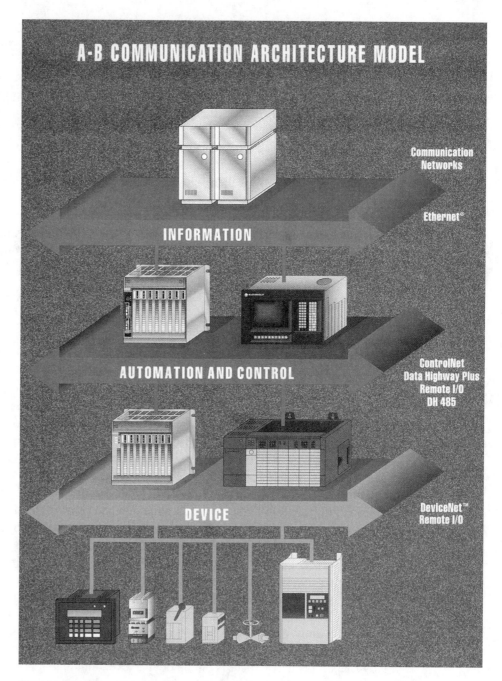

Figure 14–16 Allen-Bradley communication architecture model. (*Courtesy Rockwell Automation Inc.*)

Device Level Networks

Device level networks are becoming more prevalent in industrial control. Sometimes called *field buses* or *industrial buses*, these networks have many similarities to the conventional office computer network. Office networks allow many computers to communicate without being directly hooked to each other. Each computer has only one connection to the network bus. Networks provide several advantages: they minimize the amount of wiring needed and they allow devices such as printers to be shared. Industrial networks share some of these advantages.

Imagine a complex automated machine that has hundreds of I/O devices and no device network. Now imagine the time and expense required to connect each I/O device to the controller. This can require hundreds of wires (or more) in a complex system and can involve long runs of wire. Conduit must be bent and mounted for the vast number of wires that need to be run. Typical devices such as sensors and actuators require that two or more wires are connected at the point of use and then run to an I/O card at the control device. By contrast, in an industrial network, only a single twisted-pair wire bus needs to be run (see Figure 14-17). All devices can then connect directly to the bus.

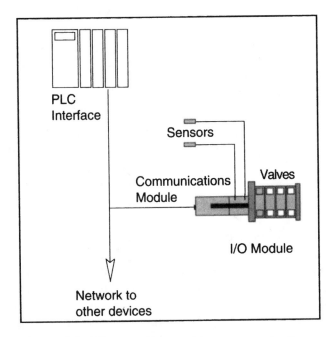

Figure 14–17 Simple bus with a communications module and an I/O module. The communications module is DeviceNet in this example. (*Courtesy Parker Hannefin Corporation.*)

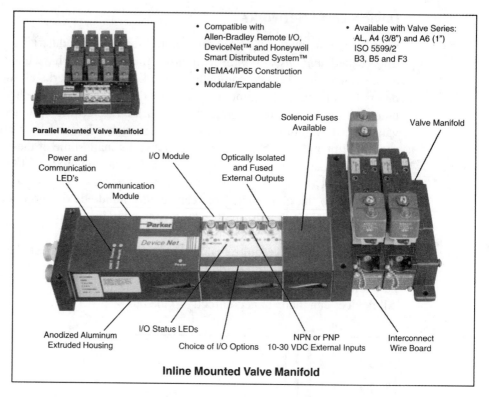

Figure 14–18   DeviceNet valve manifold. (*Courtesy Parker Hannefin Corporation.*)

Multiple devices can even share one connection with the bus in some cases. Several manufacturers have developed I/O blocks that allow multiple I/O points to share one connection to the bus. One example is the valve manifold (see Figure 14-18). This particular unit could have up to 32 valves and up to 32 I/Os with only two connections. Use of industrial networks drastically reduces wiring and installation costs.

The cost of field bus devices is higher than that of conventional devices, but these bus devices gain a cost advantage when one considers the labor cost of installation, maintenance, and troubleshooting. The cost benefit received from using distributed I/O is really in the labor saved during installation and startup. There is also a material savings from the wire that does not have to be run. There is a tremendous savings in labor considering that only a fraction of the number of connections need to be made and only a fraction of the wire needs to be run. A field bus system is also much easier to troubleshoot than a conventional system. If a problem exists, only one twisted-pair cable needs to be checked. A conventional system might require the technician to sort out hundreds of wires.

Field Devices

Sensors, valves, actuators, and starters are examples of I/O that are called *field devices*, which can be digital or analog. Industrial applications often need analog information such as temperature and pressure. These devices typically convert the analog signal to a digital form and then communicate the digital signal to a controller. They may also be able to pass other information such as piece counts, cycle times, and error codes. Devices such as drives have also gained capability. An example of a smart device is a digital motor drive to which we could send parameters and commands serially. Parameters such as acceleration and other drive parameters can be sent by a computer or PLC to alter the way a drive operates. Until recently, every device manufacturer has had different communication protocols for communication.

Figure 14-19 shows a simple industrial network with several field devices attached to it. The I/O block (I/O concentrator) and the valve manifold are used to connect multiple devices to a network.

Types of Device Level Buses

The two basic categories of device bus networks are process and device.

Device Buses Device-type buses are intended to handle the transmission of short messages, typically a few bytes in length. Process buses are capable of transmitting larger messages. Most elements in a device bus are discrete (sensors, push buttons, limit switches, etc). Many discrete buses can utilize some analog devices as well, typically those that require only a few bytes of information transmission. Examples are some temperature controllers, some motor drives, and

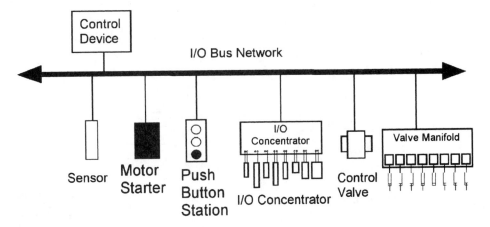

Figure 14–19 Simple industrial network.

thermocouples. Since the transmission packets are small, device buses can transmit data packets from many devices in the same amount of time it would take to transmit one large packet of data on a process bus.

Industrial buses range from simple systems that can control discrete I/O (device type bus) to buses that could be used to control large complex processes or a whole plant (process bus). See Figure 14-20 for a graph of the capabilities of several of the more common industrial buses. The left of the graph begins with simple discrete I/O control and goes up in capability to plant control. Block I/O devices such as manifolds have several valves on one block and need only two to four wires to be connected to control all of the I/O. By using an industrial bus, only two to four wires would have to be run instead of a few dozen.

Process Buses   Process buses are capable of communicating several hundred bytes of data per transmission. They are slower because of their large data packet size. Most analog control devices do not require fast response times, so most devices in a process bus network are analog. Process controllers typically are smart devices controlling analog types of variables such as flow, concentration, and temperature. These processes are typically slow to respond. Process buses transmit process parameters to process controllers.

Process Bus Standards   Two main organizations are working to establish process bus standards: the Fieldbus Foundation and the PROFIBUS (Process Field Bus) Trade Organization.

The *Fieldbus Foundation's Fieldbus* supports high-speed applications in discrete manufacturing such as motor starters, actuators, cell control, and remote I/O. Fieldbus is a new digital communications network that can be used in industry to replace the existing 4-20 milliamp analog signal. The network is a digital,

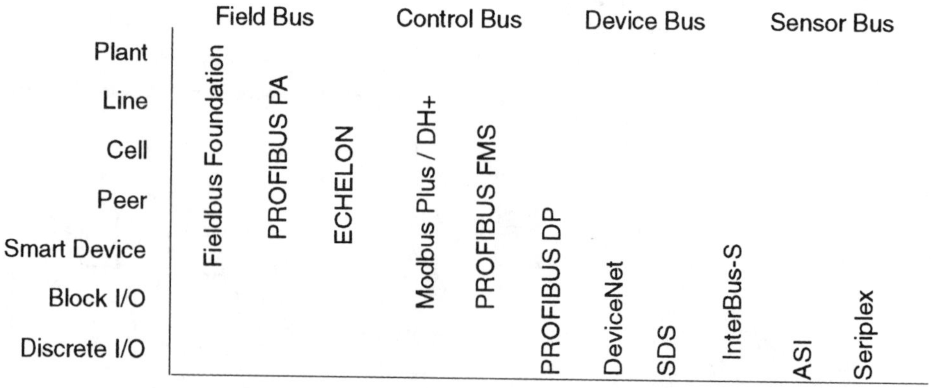

Figure 14-20   Comparison of some more common industrial buses.

bidirectional, multidrop, serial-bus communications network used to link isolated field devices, such as controllers, transducers, actuators, and sensors. Fieldbus devices can transmit and receive multivariable data and can communicate directly with each other over the bus. The ability to communicate directly between two field devices rather than having to communicate through the control system enhances the system's performance.

Each field device has low-cost computing power, making each a "smart" device. Each device is able to execute its own simple functions such as diagnostic, control, and maintenance and to provide bidirectional communication capabilities. With these devices the engineer is not only able to access the field devices but also to communicate with other field devices. In essence, Fieldbus will replace centralized control networks with distributed-control networks.

Foundation Fieldbus is an open bus technology that is compatible with Instrument Society of America (ISA) SP50 standards. It is an interoperable, bidirectional communications protocol based on the seven-layer communications model of the International Standards Organization's Open System Interconnect (OSI/ISO). The Foundation's protocol is not owned by any individual company or controlled by any nation or regulatory body. Foundation Fieldbus allows logic and control functions to be moved from host applications to field devices. The Foundation Fieldbus protocol was developed for critical applications with hazardous locations, volatile processes, and strict regulatory requirements. Each process cell requires only one wire to be run to the main cable with a varying number of cells available. The Fieldbus protocol enables multiple devices to communicate over the same pair of wires. New devices can be added to the bus without disrupting control. See Figure 14-21 for a few of the characteristics of Fieldbus.

Fieldbus eases system debugging and maintenance by allowing on-line diagnostics to be carried out on individual field devices. Operators can monitor all of the devices included in the system and their interaction.

Measurement and device values are available to all field and control devices in engineering units. This eliminates the need to convert raw data into the required units. This also frees the control system to perform other tasks.

With Fieldbus technology, field instruments can be calibrated, initialized, and operated directly over the network. This reduces time for technicians, operators, and maintenance personnel.

PROFIBUS specifies the functional and technical characteristics of a serial fieldbus. This bus interconnects digital field devices in the low (sensor/actuator level) up to the medium (cell level) performance range. The system contains master (active stations) and slave devices (simple devices such as valves, actuators, and sensors).

Slaves can respond only to received messages or send messages at the master's request. Master devices have the capability to control the bus. When the master

Network Size	Up to 64 nodes
Network Length	500 meters at 125 Kbps 250 meters at 250 Kbps 100 meters at 500 Kbps
Topology	Linear (trunkline/dropline). Power and signal on the same cable.
Physical Media	Twisted pair for signal and power
Data Packets	0 to 8 bytes
Method	Peer-to-peer with multicast (one-to-many): master/slave special case and multimaster: polled or change-of-state.
Bus Access	Carrier sense/multiple access

Figure 14–21   Characteristics of a Fieldbus network.

has the right to access the bus, it may send messages without a remote request. Figure 14-22 shows an example of a PROFIBUS network with a PC card used as a PROFIBUS master. Figure 14-23 shows a GE Fanuc communicating with PROFIBUS I/O by using a PC card.

The PROFIBUS access protocol includes the token-passing method for communication between complex stations (masters) and the master-slave method for communication between complex stations and simple peripheral devices (slaves). It is

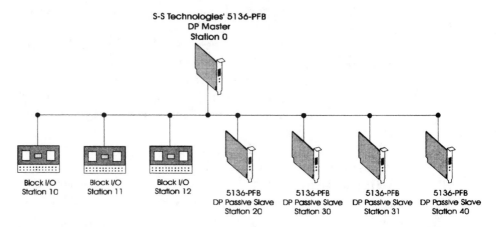

Figure 14–22   S-S Technologies PC card acts as a PROFIBUS DP master.
(*Courtesy S-S Technologies Inc.*)

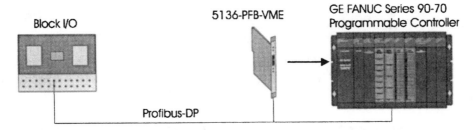

Figure 14–23 S-S Technologies PC card enables a GE Fanuc PLC to communicate with and control PROFIBUS DP I/O. (*Courtesy S-S Technologies Inc.*)

circulated in a maximum token rotation time between all masters. This maximum time is user configurable. PROFIBUS uses the token-passing method only between masters. Figure 14-24 shows some characteristics of a PROFIBUS network.

The master-slave method allows the master that currently has the token to communicate with slave devices. Each master has the ability to transmit data to the slaves and receive data from the slaves.

PROFIBUS allows the user choices of access method:

Pure master-slave system.

Pure master-master system (token passing).

System with a combination of both methods (*hybrid system*).

Network Size	Up to 64 nodes
Network Length	500 meters at 125 Kbps 250 meters at 250 Kbps 100 meters at 500 Kbps
Topology	Linear (trunkline/dropline). Power and signal on the same cable.
Physical Media	Twisted pair for signal and power
Data Packets	0 to 8 bytes
Method	Peer-to-peer with multicast (one-to-many): master/slave special case and multimaster: polled or change-of-state.
Bus Access	Carrier sense/multiple access

Figure 14–24 Characteristics of a PROFIBUS network.

Device Buses   The two categories of device buses are bitwide and bytewide. Bytewide buses can transfer 50 or more bytes of data at a time. Bitwide buses typically transfer one to eight bits of information to/from simple discrete devices. Bytewide systems are excellent for higher level communication, and bitwide systems are best for simple, physical level I/O devices such as sensors and actuators. This chapter concentrates on device level buses, DeviceNet in particular.

Actuator Sensor Interface (ASI) Bus   Actuator Sensor Interface (ASI) is a versatile, low-cost "smart" cabling solution designed specifically for use in low-level automation systems. It is easy to install, operate, and maintain; simple to reconfigure; and is low cost per node. Developed by a consortium of European companies, the technology is now owned by AS International, an independent organization. Cost savings of up to 40 percent are possible in typical automation situations. ASI technology has been submitted for approval under the proposed IEC 947 standard. It has user groups in eight European countries. See Figure 14-25 for characteristics of ASI.

In an ASI network, data and power are carried over a single cable that links up to 31 slave devices to each master and up to 124 digital input and output nodes to a PLC or computer controller. Devices such as LEDs and indicators can be powered directly because the cable can handle up to 8 amps. Higher power devices such as actuators are powered separately from a second cable. Worst-case cycle times for a full network are 5 milliseconds per slave; faster times are possible

Network Size	Up to 64 nodes
Network Length	500 meters at 125 Kbps 250 meters at 250 Kbps 100 meters at 500 Kbps
Topology	Linear (trunkline/dropline). Power and signal on the same cable.
Physical Media	Twisted pair for signal and power
Data Packets	0 to 8 bytes
Method	Peer-to-peer with multicast (one-to-many): master/slave special case and multimaster: polled or change-of-state.
Bus Access	Carrier sense/multiple access

Figure 14–25   ASI bus characteristics table.

when fewer devices are connected. The cable is mechanically polarized and cannot be connected incorrectly (that is, it is foolproof). The cable is self-healing, enabling I/O nodes to be disconnected and reconnected easily. The ASI network utilizes cyclic polling of every network participant (sensor or actuator). With 31 slaves, the cycle time is 5 milliseconds. Error detection is used to initiate a message repeat signal.

LonWorks LonWorks has been used for a wide variety of applications from very simple control tasks to very complex applications requiring thousands of devices. LonWorks networks can range in size from 2 to 32,000 devices. Figure 14-26 shows characteristics of LonWorks. It has been extensively used for building control and for slot machines, home control, petroleum refining plants, aircraft, skyscraper building control, and so on.

LonWorks networks do not have a central control or master-slave arrangement. It uses a peer-to-peer architecture. Control devices called *nodes* have their own intelligence and can communicate with each other using a common protocol. Each is able to implement the protocol and perform control functions.

Devices such as switches, sensors, relays, drives, motion detectors and other home security devices, and so on may all be nodes on the network; they typically perform simple tasks. The overall network performs a complex control application, such as controlling a manufacturing line, securing a home, and automating a building.

Network Size	Up to 64 nodes
Network Length	500 meters at 125 Kbps 250 meters at 250 Kbps 100 meters at 500 Kbps
Topology	Linear (trunkline/dropline). Power and signal on the same cable.
Physical Media	Twisted pair for signal and power
Data Packets	0 to 8 bytes
Method	Peer-to-peer with multicast (one-to-many): master/slave special case and multimaster: polled or change-of-state.
Bus Access	Carrier sense/multiple access

Figure 14–26 Characteristics of a LonWorks bus.

Controller Area Network (CAN)   Bosch originally developed CAN for the European automobile network to replace expensive wire harnesses with a low-cost network. Because of many auto safety concerns such as antilock brakes and airbags, CAN was designed to be high-speed and very dependable. All major European car manufacturers are developing models using CAN to connect various electronic components.

The Society of Automotive Engineers (SAE) developed the standard *SAE J1939*. It is designed to give "plug and play" capabilities to car owners. The standard uses extended (29-bit) identifiers to allocate CAN identifiers to different purposes.

CAN was designed to be a high-integrity serial data communications bus for real-time applications. Electronic chips were designed and built to implement the CAN standard and demand for these chips has exploded. Many millions of CAN chips are sold every year. This demand has led to the wide availability of reasonably priced CAN chips, which are typically 80 to 90 percent cheaper than chips for other networks. The CAN protocol is now being used in many industrial automation and control applications.

CAN is a broadcast bus-type system. It can operate up to 1 megabit/sec. Many controllers are available for CAN systems. Messages (or frames) variable in length are sent across the bus; they can be between 0 and 8 bytes in length. Each frame has an identifier that must be unique. This means that two nodes may not send messages with the same identifier. Data messages transmitted from a node on a CAN bus do not contain addresses of either the transmitting node or of any intended receiving node. Every node on the network receives each message and performs an acceptance test on the identifier to determine whether the message is relevant to it. If the message is relevant, the node processes it; otherwise it ignores it. This is known as *multicast*.

The node identifier also determines the priority of a message. The lower the numerical value of the identifier, the higher the priority. This allows arbitration if two (or more) nodes compete for access to the bus at the same time. The higher priority message is guaranteed to gain bus access. Lower priority messages are retransmitted in the next bus cycle or in a later bus cycle if other, higher priority messages are waiting to be sent. The two-wire bus is usually a shielded or unshielded twisted pair.

The ISO 11898 standard recommends that bus interface chips be designed so that communication can still continue (but with reduced signal to noise ratio) even if:

Either of the wires in the bus is broken.

Either wire is shorted to power.

Either wire is shorted to ground.

DeviceNET   DeviceNet is intended to be a low-cost method to connect sensors, switches, valves, bar code readers, drives, operator display panels, and so on to a simple network. The DeviceNet standard is based on the CAN chip standard. DeviceNet allows simple devices from different manufacturers to be interchanged and makes interconnectivity of more complex devices possible (see Figures 14-27 and 14-28).

In addition to reading the state of discrete devices, DeviceNet is able to report temperatures, to read the load current in a motor starter, to change the parameters of drives, and so on.

DeviceNet is an open network standard. It is a broadcast-based communications tool. Any manufacturer can participate in the Open DeviceNet Vendor Association (ODVA), Inc., an independent supplier organization that manages the DeviceNet specification and supports the worldwide growth of DeviceNet. ODVA works with vendors and provides assistance through developer tools, developer training, compliance testing, and marketing activities and publishes a DeviceNet product catalog. Figure 14-29 shows several characteristics of DeviceNet. DeviceNet supports strobed, polled, cyclic, change-of-state, and application-triggered data movement. The user can choose master/slave, multimaster, and peer-to-peer or a combination configuration depending on device capability and application requirements.

Higher priority data has the right of way. This provides inherent peer-to-peer capability. If two or more nodes try to access the network simultaneously, a bit-wise nondestructive arbitration mechanism resolves the potential conflict with no

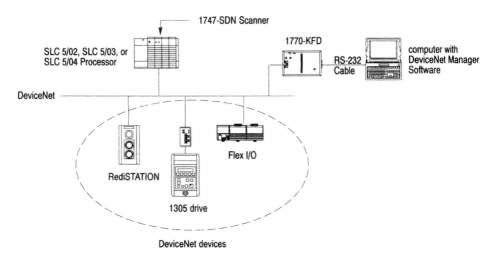

Figure 14–27   DeviceNet network. (*Courtesy Rockwell Automation Inc.*)

Figure 14–28 DeviceNet network. (*Courtesy S-S Technologies Inc.*)

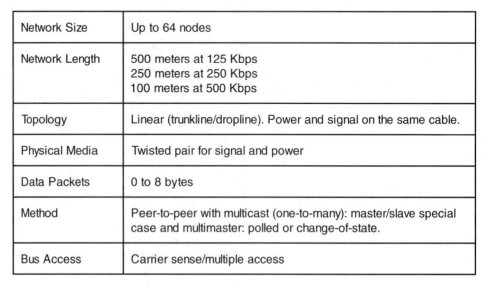

Network Size	Up to 64 nodes
Network Length	500 meters at 125 Kbps 250 meters at 250 Kbps 100 meters at 500 Kbps
Topology	Linear (trunkline/dropline). Power and signal on the same cable.
Physical Media	Twisted pair for signal and power
Data Packets	0 to 8 bytes
Method	Peer-to-peer with multicast (one-to-many): master/slave special case and multimaster: polled or change-of-state.
Bus Access	Carrier sense/multiple access

Figure 14–29 DeviceNet characteristics.

loss of data or bandwidth. By comparison, Ethernet uses collision detectors, which result in loss of data and bandwidth because both nodes have to back off and resend their data.

Peer-to-peer data exchange means that any DeviceNet product can produce and consume messages. *Master/slave operation* is defined as a proper subset of peer-to-peer exchange. A DeviceNet product may behave as a client or a server or both. A DeviceNet network may have up to 64 media access control identifiers or MAC IDs (node addresses). Each node can support an infinite number of I/O. Typical I/O counts for pneumatic valve actuators are 16 or 32.

Change-of-State and Cyclic Transmission *Change-of-state* means that a device reports its data only when the data change. To be sure the consuming device knows that the producer is still alive and active, DeviceNet provides an adjustable, background heartbeat rate. Devices send data whenever their data changes or the heartbeat timer expires. This keeps the connection alive and lets the consumer know that the data source are still alive and active. The heartbeat timing prevents talkative nodes from dominating the network. The device generates the heartbeat, which frees the controller from having to send a nuisance request periodically just to make sure that the device is still there.

The cyclic transmission method can reduce unnecessary traffic on a network. For example, instead of scanning temperature many times every second, devices can be set up to report their data on a regular basis that is adequate to monitor any change.

Device Profiles The DeviceNet specification promotes interoperability of devices by specifying standard device models. All devices of the same model must support common identity and communication status data. Device-specific data are contained in device profiles that are defined for various types of devices. This ensures that devices from multiple manufacturers, such as push buttons, valves, and starters are interchangeable. Manufacturers can offer extended capabilities for their devices, but the base functionality must be the same.

Control Level Communications

Communications at the control level occur between control-type devices including PLCs, robots, CNC controllers, and so on.

Control Level Networking The need for standardization of control level communications so that control equipment from various manufacturers can communicate has existed for some time. One of the emerging standards for control level networking is ControlNet.

ControlNet™   ControlNet is a high-speed, deterministic network developed by Allen-Bradley for transmitting critical automation and control information. It has very high throughput, 5 megabit/sec for I/O, PLC interlocking, peer-to-peer messaging, and programming. ControlNet can perform multiple functions including multiple PLCs, man/machine interfaces, network access by PC for programming and troubleshooting from any node. Its capability to perform these tasks can reduce the need for multiple networks for integration. See Figure 14-30 for an example of integrating various networks.

ControlNet is compatible with Rockwell Automation PLCs, I/O, and software. It supports bus, star, or tree topologies. It utilizes RG6 cable, which is identical to cable television cable. This means that taps, cable, and connectors are all easy to obtain and very reasonable in price. ControlNet also has a redundant cabling option (see Figure 14-31), meaning that two separate cables can be installed to guard against failures such as cut cables, loose connectors, or noise. This is called redundant cabling. See Figure 14-31 for some types of computers, controllers and wiring that can be integrated.

Ethernet   Ethernet, which is the standard for enterprise level communications, is making inroads in control level networking. It is the most common popular personal computer, because it is widely used for computer networking and is

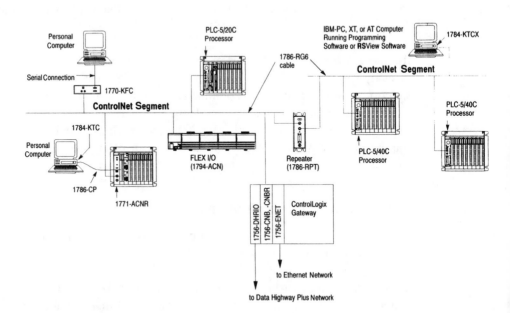

Figure 14–30   Integration of various networks. (*Courtesy Rockwell Automation Inc.*)

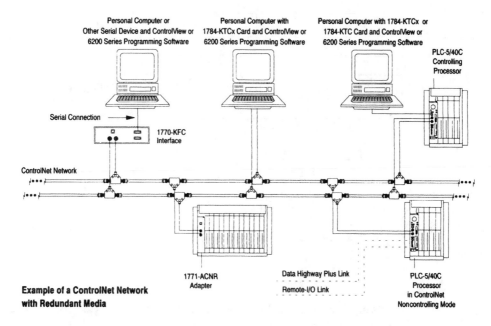

Figure 14–31 Redundant cabling. (*Courtesy Rockwell Automation Inc.*)

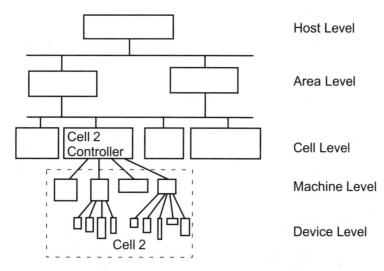

Figure 14–32 Typical enterprise level communication control, also called host-level control.

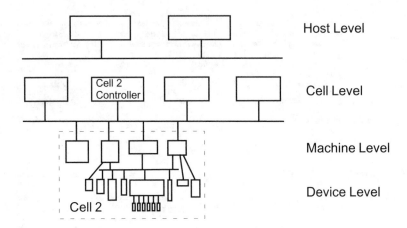

Figure 14–33 Networking example.

relatively inexpensive to implement. It is based on CSMA/CD access methods and bus structure. Increasing its speed may make Ethernet an even more popular choice. Most controllers can now be purchased with Ethernet capability.

Enterprise Level Communications

Enterprise level communications seems to have adopted Ethernet for its standard. It meets the needs of the enterprise level very nicely. Robots, PLCs, and many other devices are available with Ethernet capability. They can be plugged right into a plant's Ethernet network for communication.

The rapid advancements in computers and networking are creating vast changes in industrial communication hierarchies. Figure 14-32 is an example of enterprise level networking. Some field devices are connected to a bus type network at the device level in Figure 14-33.

QUESTIONS

1. Describe the term *serial communication*.
2. Describe the term *asynchronous*.
3. Explain the term *SCADA*.
4. Complete the following table.

	Speed	Length Limit	Balanced or Unbalanced	Special Characteristics
RS-232				
RS-422				
RS-423				
RS-485				

5. Explain industrial networks and why they are becoming more prevalent.

6. What are some of the differences between a device bus and a process bus?

7. What is a field device?

8. Describe how the capabilities of field devices are changing.

9. Where are device buses appropriate?

10. Where are process buses appropriate?

11. Why is the CAN system so important?

12. Describe the DeviceNET standard.

13. Describe the purpose of Fieldbus.

14. Describe the purpose of ControlNet.

Computer Numerical Control (CNC) Machines

15

This chapter will introduce the reader to the concepts of CNC machining, servo control, and CNC programming. Programming and operations of CNC machines will be covered.

OBJECTIVES

Upon completion of this chapter you will be able to:

1. Explain the terms *closed-loop servo, feedback, position* and *velocity loops,* M-codes, and G-codes.
2. Explain how a servo system is homed.
3. Explain the parts of a program.
4. Arrange and explain blocks of information.
5. Describe preparatory and miscellaneous functions.
6. Describe the purpose of tool height and tool diameter offsets.
7. Write a CNC program.

The trend toward automation of production equipment is putting great demands on people. Since the early 1970s, manufacturers have worked to increase productivity, quality, process capability, reliability, and flexibility. Manufacturers began to realize that they had to be competitive with foreign manufacturers if they were to survive. This meant using technologies that could improve quality and productivity.

COMPUTER NUMERICAL CONTROL (CNC) MACHINES

The improvements in computers have dramatically increased the use of computer numerical control (CNC) machines. The most common of these CNC machine tools are *turning* and *machining centers*, which together make up more than half of the numerical control machines on the market. Other types include ram and wire feed-type electrical discharge machines (EDM); flame cutting, plasma-cutting, and water jet–cutting machines; and coordinate measuring machines. Although each machine works somewhat differently, the basic method of initiating for machine action remains the same. The standards and codes developed in the early years of numerical control still apply today.

CNC Turning Center

A CNC turning center is a computer-controlled lathe (see Figure 15-1). CNC turning centers are used to produce cylindrical parts such as shafts, axels, pins and so on.

Vertical Machining Centers

Vertical machining centers are vertical milling machines that use CNC positioning and automatic tool changers to produce complex machine parts in one set up (see Figure 15-2).

Horizontal Machining Centers

Horizontal machining centers are horizontal milling machines that are numerically controlled. They are equipped with automatic tool changers and a variety of other features to increase their versatility and production capabilities.

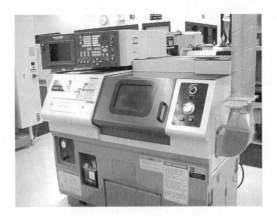

Figure 15–1   CNC turning center.

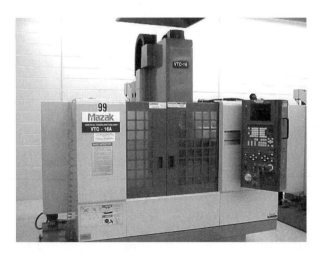

Figure 15–2   CNC vertical machining center.

Machine Tool Axes

CNC machines can also be classified according to the number of axes, or directions of motion, that they are capable of. Milling machines or machining centers usually have 2.5, 3.0, 4.0, or 5.0 or more axes with the three and four axes being the most common. The four and five axes machines incorporate a rotary table of some sort to be capable of continuous motion in all axes simultaneously. Lathes and turning centers generally have between two and four axes. The standard configuration consists of a two-axis lathe with one axis parallel to the spindle and one axis perpendicular to the spindle.

Components of Numerical Control Machines

The numerical control machine can be divided into three basic parts: control, which processes the commands from the input media; the drive mechanisms; and the machine itself.

CNC Controls

CNC controls can store loaded programs and allow their editing. Machine controls today are very powerful and flexible. Many of a machine's operating characteristics can be changed to operate as the operator wants it to. Each CDC machine can be adapted to the needs of a particular job. Simply put, the computer controls of a CNC machine tool are programmed instead of hard wired. The newest computer controls are PC based that can easily be put on the company's Ethernet computer network.

Drive Mechanisms

Servo Motors and Closed-Loop Systems   Servo motors permit untended operation. Closed-loop control makes sure that the machine actually does what the program told it to do. The increasing capability and speed of computers make it possible to continuously monitor the machine's position and velocity while it is operating. For example, if the machine is told to move 10 inches at a feed rate of 5 inches per minute, the computer constantly monitors the axis as it moves to be sure that it is properly executing the move. If the machine should run into an obstruction, the computer knows instantly that it should be moving but it is not. The computer then stops the motion and signals an error condition, usually before serious damage is done to the machine. The advantages of the servo motor include motor and feedback mechanism in one housing; increased travel and spindle speeds; and increased accuracy and repeatability.

See Figure 15-3 for an example of one complete axis of motion for a CNC machine, a robot, or any other servo-controlled machine.

Velocity Loop   A *velocity loop* consists of a motor drive, a motor, and a tachometer. The drive needs an input called a velocity reference command to tell it in which direction and how fast to run. The drive receives the velocity reference command from the CNC control. As the motor begins to turn, the tachometer, which is really a generator, turns. The faster it turns, the more voltage it outputs. It is mechanically attached to the motor shaft so that it turns at the same speed as the motor. The voltage from the tachometer, called *feedback*, is fed back to the drive. The drive has a circuit called a *comparator* (or *summing junction*) that compares the command

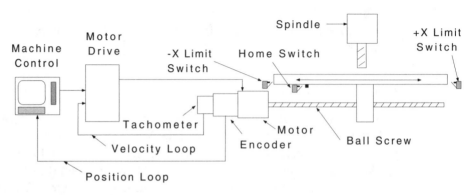

Figure 15–3   One axis of a closed-loop servo of a milling machine table.

from the CNC control and the voltage from the tachometer (feedback on the actual speed) and outputs the difference between the two, called the *error*. The feedback from the tachometer is connected so that it will be the opposite polarity of the command from the CNC control. Thus, the drive outputs the difference (or error) of the command and feedback. This ensures that the drive maintains almost perfect velocity. Beginning some heavy cuts in tool steel tends to slow the motor down, which slows down the tachometer, resulting outputs of less voltage. The drive then sums the command with the smaller feedback signal and detects the larger error. The drive then outputs more voltage (current) so that the motor speeds up and runs at the correct speed. This all happens instantaneously, so the speed really never varies. If the load became too great, the drive is protected by a current limit, which would stop the drive, probably before any damage is done. The drive then must be reset.

Position Loop   The *position loop* is closed by an *encoder* device, which has a disk with three rings around it (see Figure 15-4). Each ring consists of many slits. A typical encoder might have 1000 slits in the first ring, 1000 in the second, and only one in the third. A light is shined through each of these rings. Three light receivers, one for each ring, are on the other side of the disk. The encoder passes light to the receiver when the slit lines up with the light and receiver, creating pulses. The pulses are sent back to the CNC control, which knows that there are 1000 pulses per revolution; by counting the pulses, it knows the encoder's position exactly. The CNC control knows when the pulse of the first and second rings begins and ends, so it can actually increase the resolution.

Homing   When a machine is initially started, it needs to be "homed" because it has no idea where each axis is located. *Homing* moves each axis to a known position where each axis is initialized. It is crucial that a machine home at the exact

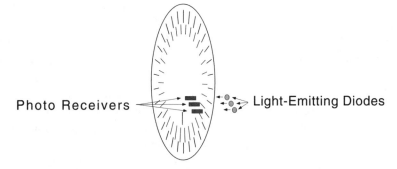

Photo Receivers   Light-Emitting Diodes

Figure 15–4   Typical encoder. The light-emitting diodes and the light breaking receivers. The CNC control knows which direction the machine is moving by seeing whether the first pulse comes from the first ring or the second ring.

same position each time so that programs and fixture locations can be accurate and consistent. The procedure is very simple. The operator moves each axis to a safe location and then sets a home key for each axis. A safe location has no fixtures, parts, or other tooling that the machine will run into as it homes. Let's examine what happens when one axis is homed. Study the position loop in Figure 15-3 and the table switches. The machine table can contact three switches: a –X limit switch, a +X limit switch, and a home switch. The limit switches are for overtravel protection to disable the motor drive if the table ever moves far enough to contact them. If the table contacts the –X limit switch, for example, the motor drive is disabled, preventing any further movement in the –X direction. The control goes into an error condition, and the operator must reset the error and move the axis in the +X direction until the table is off the –X limit switch. The same is true if the table contacts the +X switch. The home switch roughly positions the table during the homing routine. The operator moves the table to a safe position and touches the home key for the X axis (in this example). The table begins to slowly move toward the home switch. When the table contacts the switch, a signal is sent to the CNC control. It senses that the switch has closed and then reverses the motor and very slowly turns it. The CNC control monitors the encoder pulses now until it sees the home pulse. When the CNC control sees the home pulse from the encoder, it initializes its position. It now knows exactly where this axis is. So the home switch ensures that the table is close to position and in the right revolution of the encoder. The encoder pulse then ensures that the machine is exactly in position. Each axis homes in the same manner. Most machines also have software axis limits in computer memory. The software limits are set up to be activated before the actual limit switches. This provides extra protection.

CNC Machine

The machine tool itself has evolved with the computer control and the drive systems. In the beginning, the majority of numerical control machines were conventional machine tools married to a control (often called *retrofit*). Early numerical control machines did not have computers but used punch cards as "code" to control the machines axes and functions.

Modern machine tools have been completely redesigned for numerical control machining. These machines bear little resemblance to their conventional counterparts. Requirements of CNC machine tools include rigidity, rapid mechanical response, low inertia of moving parts, and low friction.

Tool Changing Tool changers on CNC machines allow automatic tool changing without operator intervention. Milling machining centers can have from 8 to 100 tools; lathe or turning centers typically have 6 to 12 tools that are indexed automatically. Speed in tool changing is an important factor for machines used in production environments.

Programming and Input Media

Numerical control programs may be produced manually and entered into a computer or produced directly in the machine. One method involves the use of a computer-aided parts programming system (CAPP). The system allows the programmer to describe the part and the tool paths using a special language or screen prompts. The computer then processes the tool paths into numerical control language that is then loaded into the control.

Another method of programming generates the program using the conversational or symbolic programming systems built into some machine controls. In many systems, the program appears as machine language (i.e., G-Code and M-Code programs). For this reason, programmers and operators need a strong background in machine code language.

AXIS AND COORDINATE SYSTEMS

To fully understand numerical control programming you must understand axes and coordinates. You could describe any part to someone else by its geometry. For example, the part is a 4 × 6 inch rectangle. If the part has holes, they can be described by their position. Any point on a machined part can be described in terms of its position. The system that allows us to do this is called the *Cartesian coordinate* or *rectangular coordinate system* developed by a French mathematician, Renee Descartes.

Consider a single-axis coordinate line called the x-axis (Figure 15-5). A vertical line perpendicular to the x-axis crosses the x-axis at the zero point, also called the *origin* (Figure 15-6). Points are described by their distance along the axis and by their direction from the origin by a plus (+) or minus (−) sign. We could tell anyone where a particular position is on a line by giving its position from zero, called a *position's coordinate*.

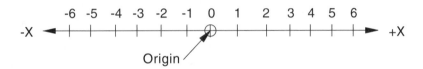

Figure 15–5 Example of Cartesian coordinate positioning, or rectangular coordinates.

Figure 15–6 Dual-axis coordinate grid.

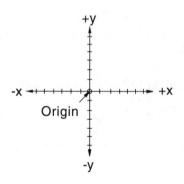

Quadrants

The axes divide a work envelope into four sections called *quadrants* (see Figure 15-7). The quadrants are numbered counterclockwise starting from the upper right. Points in the upper right, quadrant 1, have positive x and y (+x, +y) values. Points in quadrant 2 have a negative x (−x) and a positive y (+y) value. Points in quadrant 3 have a negative x and y (−x, −y) value. Points in quadrant 4 have a positive x (+x) and a negative y (−y) value.

See Figure 15-8 for a point in the Cartesian coordinate system whose coordinates are X3, Y3. Only one point matches these coordinates.

Figure 15-9 shows a three-axis system. A third axis represents depth. We describe the location of a point where a hole is to be identified by its x and y coordinates. We would use a z value to represent the depth of the hole. The z-axis is added perpendicular to the x- and y-axes.

Consider a vertical milling machine (see Figure 15-10). The x-axis is the table movement right and left as you face the machine. The y-axis is the table movement toward and away from you. The z-axis is the spindle up and down. A move toward

Figure 15–7 Four quadrants in the Cartesian coordinate system.

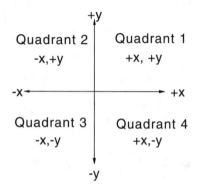

Figure 15–8 Point identified by its coordinates in the xy plane.

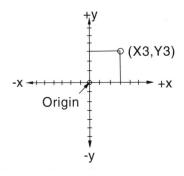

Figure 15–9 Three-axis coordinate grid.

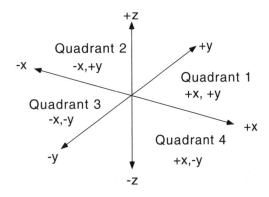

Figure 15–10 Typical milling machine configuration illustrating the x-, y-, and z-axes of motion.

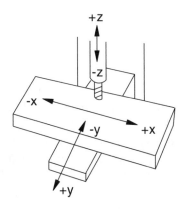

Figure 15–11 Typical lathe configuration illustrating the x- and z-axes.

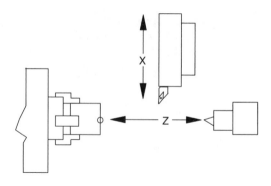

the work is a negative z ($-z$) move. A move up this axis is a positive z ($+z$) move. Milling machines use all three axes. The x-axis usually has the longest travel. On a common vertical milling machine, the x-axis moves to the operator's left and right. The y-axis moves toward and away from the operator; it usually has the shortest travel. The z-axis always denotes movement parallel to the spindle axis. The z-axis on the vertical mill is an up and down movement. Toward the work is a negative z move. Remember to think of these moves in relation to the spindle.

Lathes or turning centers typically use only the x- and z-axes (See Figure 15-11). The z denotes movement parallel to the spindle axis that controls the lengths of parts or shoulders. The x-axis is perpendicular to the spindle and controls the diameters of the parts.

ABSOLUTE AND INCREMENTAL PROGRAMMING

Absolute and incremental programming coordinates specify the relative tool-moving position with respect to the program zero or to the tool-moving distance from its current position. Computer programs can be written in absolute or incremental programming format or a mix of the two. *Absolute coordinates* are located in relation to the program zero. *Absolute programming* specifies a position or an end point from the work piece's coordinate zero. It is an absolute position.

In incremental coordinate positioning, the spindle's present position becomes the program start position. From the start point, point 1 is one position to the right on the x-axis ($x + 1$) and six positions up on the y-axis ($y + 6$). Now let us position point 2. Point 1 now becomes our start point. Point 2 is five positions to the right of our present position on the x-axis ($x + 5$) and two positions down on the y-axis ($y - 2$). Remember that moves down or moves to the left are negative moves in incremental programming.

Incremental programming specifies the movement or distance from the machine's is current point. Remember that a move to the right or up from this position is always a positive move ($+$) and a move to the left, or down, is always a negative move ($-$). With an incremental move we specify how far and in what direction we want the machine to move.

Absolute positioning systems have a major advantage over incremental positioning. A mistake made during programming using absolute positioning is isolated to the one location. A positioning error made using incremental positioning affects all future positions.

FUNDAMENTALS OF NUMERICAL CONTROL PROGRAMMING

Numerical control programming takes the information used to do manual machining and converts it to a language that the machine understands. This book uses the word address programming language. It is the most common machine language in use today.

Word Address Programming

The word address format programming for numerical control programming precisely controls machine movement and function through the use of short sentencelike commands. These commands consist of addresses, words, and characters. A CNC machine must do the same things the operator of a manual machine does to produce a part.

The process to perform a simple operation on a manual milling machine and convert it to word address is as follows. Turn the spindle on in a clockwise direction at a spindle revolutions per minute of 600. The operator sets the speed and turns the spindle in a clockwise direction. The command for a CNC machine is N0010 M03 S600; N0010 is a line number. M03 tells the spindle to start in a clockwise direction. S600 tells the spindle how fast to turn. The semicolon lets the control know that you have completed this command block and that it is time to move to the next command. A full command or block of information is made up of addresses and characters.

Letter Address Commands The command block controls the machine tool through the use of letter address commands. The following are abbreviated descriptions of the most common letter address commands:

N is the letter address for the line number or sequence number of the block.

G code is a preparatory function that sets up the mode in which the rest of the operation(s) are to be executed.

F is the feed rate of the controlled axis.

S is a spindle speed setting.

T is the letter address for a tool call.

M is a miscellaneous function such as coolant on/off, spindle forward, and so on.

H,D are auxiliary letter address codes for tool offsets storage.

Part Program   A part program is a series of command blocks that execute motions and machine functions in order to manufacture a part. We use the example of a simple program that cuts around the outside of a 3 inch × 4 inch block with a 1/2 inch diameter endmill (see Figure 15-12). The coordinate dimensions reflect the position of the center of the spindle, so we must accommodate for the size of the cutter.

Following is the program to produce the part shown in Figure 15-12.

N0010 G00 X-1.00 Y-1.00 (point 1)

N0020 G01 X-.25 Y-.25 F10.0 (point 2)

N0030 G01 Y3.25 (point 3)

N0040 G01 X4.25 (point 4)

N0050 G01 Y-.25 (point 5)

N0060 G01 X-.50 (point 6)

The first line, N0010, rapidly positions the tool just off of the lower left-hand corner of the part. In the second line, N0020, the G01 moves the tool to a position that is 1 tool radius value to the left of the side of the part (X-.25). The tool is feeding at 10 inches per minute (F10). It is now aligned with the left-hand corner of the part. This prepares for time to cut the left side of the part. Line 30 cuts the left side of the part and positions the spindle center past the top of the part by the

Figure 15–12   First part of program to be programmed.

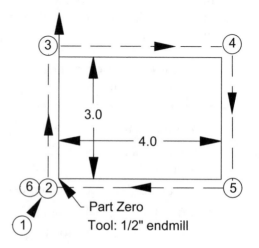

tool radius (Y3.25). This positions the edge of the tool for the cut across the top of the part.

Line 40 cuts the top of the part. The feed rate is still 10 inches per minute because we have not changed it since line number 20. The spindle center is now positioned .25 to the right of the part. This prepares for cutting the right-hand side of the part. Line 50 positions the tool to point 5. The right-hand side of the part is now complete. The tool center is also positioned 1 tool radius below the bottom of the part, ready to cut. Line 60 cuts the bottom of the part to size and moves the tool completely off of the part (X.50).

Now, let's consider the individual parts of a word address part program in-depth.

Part Datum (Workpiece Zero) Location   To program a part, the programmer must determine where the part datum zero or workpiece should be located. The part datum is a feature of the part that holds the majority of the dimensions of the part. All dimensions of the part shown in Figure 15-12 come from the lower left-hand corner, which was the logical choice for the zero point. Good programmers choose a part feature that is easy to use.

Sequence Numbers (NXXXX)   Sequence numbers identify blocks of information within the program. In most cases sequence numbers are not required because the machine will execute blocks of information in the order in which it reads them. Sequence numbers can be very helpful, however, in identifying problems.

The machine controller can be commanded to find blocks of information by their sequence numbers. They are needed in some canned cycles and will be covered in later sections of the book.

Preparatory Functions (G Codes)   Preparatory functions set the control for various machine movements such as linear interpolation (G01) and rapid traverse (G00). A two-digit number preceded by a G code determines the machining mode of the block or line that it is in. G codes or preparatory functions fall into two categories: modal or nonmodal. Nonmodal or "one-shot" G codes are command codes that are active only in the block in which they are specified. Modal G codes are command codes that remain active until another G code in the same group overrides it. If five command lines were all linear feed moves, a G01 is necessary only in the first line. The next four lines would use the previous G01 code.

Spindle Control Functions (S)   Spindle speeds are controlled with an S letter address followed by up to four digits. When programming the machining center, the spindle speed is programmed in revolutions per minute (rpm). A spindle speed of 600 rpm is programmed S600.

Miscellaneous Functions (M)   Miscellaneous functions or M codes perform miscellaneous machine functions such as tool changes, coolant control, and spindle operations. An M code is a two- or three-digit numerical value preceded by a letter address M. M codes, like G codes, can be modal or nonmodal.

Tool Call (T)   The tool call block is fairly straight forward, although the machining center differs slightly from the turning center. The tool call always starts with a T and then the tool number (T02). A tool change on a machining center requires a miscellaneous code of M06. The control is then told which tool to change to (T02). A typical tool change block for a machining center is N0010 M06 T02. On the turning center, the tool call also starts with a T and then the tool number (T02), and then the tool offset number is added. T0202 is the tool call for tool number 02 with an offset of number 02. It is written 02 because typically more than 10 tools and 10 offsets are available, for example, T1212 (tool 12, offset 12). The offset gives the operator the ability to correct any errors in the size of the part. It is not necessary to use an M06 on the turning center to do a tool change; in fact, an M06 on the turning center usually unclamps the chuck.

Axes Words (x, y, z)   Typical machining centers have three axes of motion: x, y, and z. The letter address may be proceeded by a direction sign ($-$). A simple command block to rapidly position the tool of the milling machine to within 1 inch of the work zero is carried out by this line: N0010 G00 Z1.00.

Types of Motion

Tool or table motion is controlled in three ways: rapid positioning, linear feed, and circular feed.

Rapid Traverse Positioning (G00)   A rapid positioning block consists of a preparatory or G code and the coordinate of the desired position. A rapid move to a location of x10, y5, and z1 is programmed as G00 X10.0 Y5.0 Z1.0. After this block is commanded, the machine moves at a rapid traverse rate to this position, moving all of the axes commanded simultaneously. The rapid traverse rate for each machine is different but normally ranges from 100 inches per minute to 600 or more inches per minute. The rapid traverse rate can usually be overridden using the rapid traverse override switch located on the control.

Linear Feed Mode (G01)   A linear feed mode G01 moves the tool to a commanded position in a straight line at a specific feed rate. The feed rate is the speed at which the machine axes move. Linear feed blocks are normally cutting blocks.

Figure 15–13 Linear interpolation example.

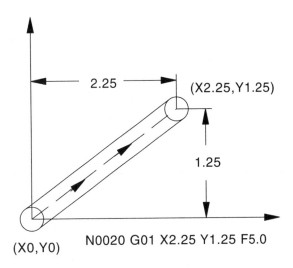

The rate at which the metal is removed is controlled through an F, or feed, rate code. Machining centers, like manual milling machines, use feed rates in inches per minute (ipm). Turning centers are typically programmed in inches per revolution of the spindle (ipr). To make a straight line cutting motion on a machining center, the block of information is: G01 X10.00 F10.00. The tool would move to an x-axis position of 10.00 inches at a feed rate of 10 inches per minute. Straight line moves can also be angular. CNC machine controls are capable of making simultaneous axes moves (see Figure 15-13).

G01 Linear Interpolation Example The G00, G01, and F codes are all modal. Modal commands are active unless changed by another preparatory code. To program a series of straight-line moves, only the G01 and the feed rate are needed in the first line. The following lines are controlled by the previous G01 and feed rate. To change to a rapid positioning mode, a G00 is used at the beginning of that line.

The sample part in Figure 15-14 incorporates some basic programming procedures. The machining procedure is to mill around the outside of the part, .500 deep, with a .50 inch endmill at a feed rate of 5 inches per minute. The first portion of the program sets up the control using preliminary procedures. The second part of the program sets the tool call. The third step sets the WPC or workpiece zero point. The fourth portion of the program starts the spindle and sets the number of revolutions per minute. The fifth part of the program sets the rapid position close to the part and starts linear cutting moves. After the profile of the part has been cut, the machine returns to the home position and ends the program.

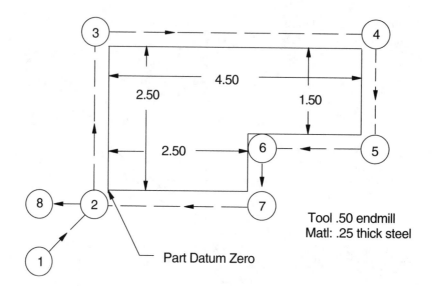

Figure 15–14 Sample part.

Following is an example of a basic contour program.

N0010 G70 G90; (Inch programming, absolute programming)

N0020 M06 T02; (Tool change, tool 2)

N0030 G54 X-10.250 Y-8.750 Z-7.525; (The G54 and the X, Y, and Z values tell the control where the part is on the table)

N0040 M03 S800; (Spindle start clockwise, 800 revolutions per minute)

N0050 G00 X-1.00 Y-1.00; (Rapid to position 1 just off the lower left corner of the part)

N0060 G00 Z.100; (Rapid down to .100 clearance above the part)

N0070 G01 Z-.50 F5.0; (Feed down to depth at 5 inches per minute)

N0080 G01 X-.25 Y-.25; (Feed to position 2)

N0090 G01 Y2.750; (Feed to position 3)

N0100 G01 X4.75; (Feed to position 4)

N0110 G01 Y.75; (Feed to position 5)

N0120 G01 X2.75; (Feed to position 6)

N0130 G01 Y-.25; (Feed to position 7)

N0140 G01 X-1.00; (Feed to position 8)

N0150 G28; (Return all axes to home position)

N0160 M06 T0; (Tool change, puts the tool away)

N0170 M30; (Rewinds the program, resets the control, and ends the program)

Programming Procedures

All CNC programs, whether for a machining center or a turning center, follow a common general procedure. The procedure is as follows:

1. Start up or preliminary procedures
2. Tool call
3. Workpiece location block
4. Spindle speed control
5. Tool motion blocks
6. Home return

Tool Changes A miscellaneous code M06 calls for a tool change, and the T code tells the control which tool changer pocket the tool is in.

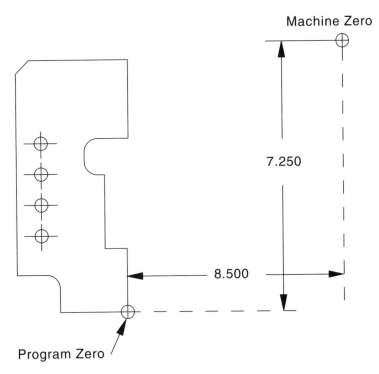

Figure 15–15 Comparison of machine zero and program zero.

Workpiece Coordinate Setting Workpiece coordinate (WPC) involves setting two important factors. The first factor is where on the part datum the part is located. The other is where on the machine table that part datum is. The WPC tells the machine control the position of the part datum.

The workpiece coordinate setting references the machine to the part location (see Figure 15-15). Although the workpiece datum point may be located at the corner or any other part of the workpiece, the control must be told where this point is on the machine table. The technique for locating the workpiece zero varies for each machine tool. Some controls use a button for setting the zero point. The setup person or operator uses the jog buttons to position the spindle center over the part zero and then presses the zero set or zero shift button to set the coordinate system to zero. On other types of controls, the WPC is set with a G code. Systems of this type use a G54 or a G92 followed by dimensions (see Figure 15-16 for an example of a G92).

The G54 workpiece coordinate setting is an absolute coordinate position from the machine home position. It is written with negative values because the part is located to the left and down on the x- and y-axes. To locate this position, the op-

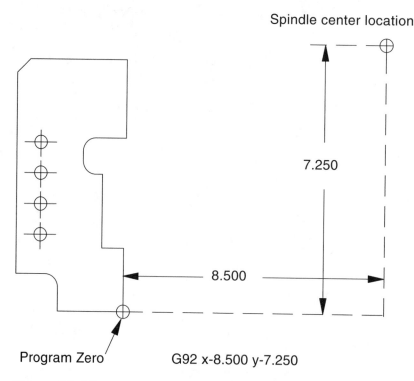

Spindle center location

7.250

8.500

Program Zero   G92 x-8.500 y-7.250

Figure 15–16 G92 workpiece coordinate example.

erator positions the center of spindle directly over the part datum using an edge finder or probe and then notes the machine position. The coordinates of this position are placed in the G50 or G54 line. A typical workpiece coordinate setting of this type is written N0010 G54 x-8.500 y-7.250 z-15.765.

The G92 workpiece coordinate is the incremental distance from the workpiece zero to the center of the spindle. When the G92 is called, the center of the spindle must be in the preprogrammed position. If it is not, the control will start machining at the wrong position. Let's say, for example, that the center of the spindle, at the home position, is 10 inches to the right on the x-axis, 5 inches in on the y-axis, and 8 inches up on the z-axis; the G92 is written G92 x10.00 y5.00 z8.00. If the center of the spindle were located any other distance away from the part zero when this G92 were called, the tool would cut the part in the wrong location. For this reason, using a G92 can be very dangerous! The G54 type of WPC setting is much safer than a G92. No matter where the spindle is when the G54 is called, the control knows exactly where part zero is located because it is an absolute position, not an incremental distance (see Figure 15-17).

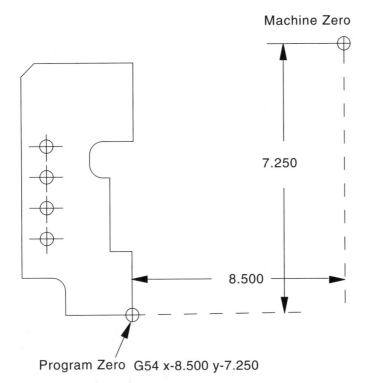

Figure 15–17   G54 workpiece coordinate example.

The G92 workpiece coordinate setting is the incremental distance from the part zero to the location of the center of the machine spindle. It is important to remember that a G54 or G92 will not move the machine tool to this point but merely lets the control know where the part is.

Spindle Start Block   The spindle is controlled by two codes: M03 tells the spindle to start in a clockwise direction, and S1000 tells it how fast to turn.

Tool Motion Blocks   The tool motion blocks are the body of the program. The tool is positioned and the cutting takes place in these blocks.

Home Return   The tool needs to be returned to home whenever a tool change takes place. Some machine controls use a G28 command to return to home; other controls return to home automatically when an M06 tool change is commanded. When a tool change or home return is commanded, it is important to note how the tool gets there. Does the z-axis move straight up to a clearance position first, or do all of the axes move simultaneously toward the home position?

Program End Blocks   There are a number of different ways to end the program. Some controls require turning off the coolant and the spindle with individual miscellaneous function codes. Other controls end the program, rewind the program, and turn off miscellaneous functions with an M30 code.

Incremental Positioning   Absolute positioning requires the coordinates of the part program to be related to an absolute zero point. Incremental positioning, also known as *point-to-point positioning*, defines the coordinates of the part in relationship to the final position of the previous move. Incremental positioning can be very useful when programming a series of holes that are incrementally located on the part print. The current point location is the datum for the next coordinate position.

This type of programming can apply to the whole part or just certain sections of the program. Incremental positioning is performed by the use of a G91 preparatory code. To switch back to absolute programming at any point in the program, a G90 code is used.

Circular Interpolation   Up to this point we have discussed only straight-line moves. If a computer numerical control machine were capable of only straight-line moves, its use would be very limited. One of the most important features of a CNC machine is its ability to do circular cutting motions. CNC machines are capable of cutting any arc of a specified radius value. *Arc* or *radius cutting* is known as *circular interpolation*. It is performed with the use of G02 or G03 preparatory code. To cut an arc, the programmer needs to follow a very specific procedure.

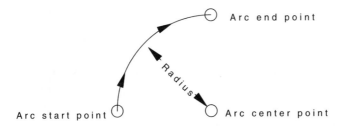

Figure 15–18   Arc start point, arc end point, and arc center point.

The critical information needed to cut an arc are the arc start point, arc direction, arc end point, and arc center point location. When starting to cut an arc, the tool is already positioned to the arc's start point. Next, the control must be told the direction of the arc, clockwise or counterclockwise. The third piece of information the control needs is the end point of the arc. The last piece of information the control needs to know is either the position of the arc center or the radius value of the arc.

Arc Start Point   The arc start point is the coordinate location of the start point of the arc (see Figure 15-18). The tool is moved to the arc start point in the line prior to the arc generation line. Simply stated, the start point of the arc is the point where you are presently when you want to generate an arc.

Arc Direction (G02, G03)   Circular interpolation can be performed in two directions, clockwise or counterclockwise (see Figure 15-19). Two G codes specify arc direction. The G02 code is used for circular interpolation in a clockwise direction, and the G03 code is used for circular interpolation in a counterclockwise direction. Both G02 and G03 codes, which are modal, are controlled by a feed rate (F) code, just like a G01.

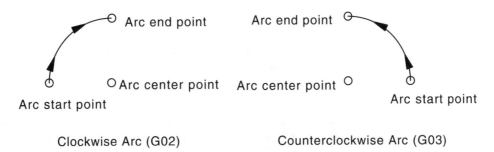

Clockwise Arc (G02)   Counterclockwise Arc (G03)

Figure 15–19   Clockwise and counterclockwise arcs.

Arc End point The arc end point is simply the x and y position for the end point of the arc.

Arc Center Points To generate an arc path, the controller must know where the center of the arc is (see Figure 15-20). The arc center position must be programmed; x and y were already used to specify the arc endpoint. We use I and J to describe the x and y arc center point. I = x-axis coordinate of an arc center point and J = y-axis coordinate of an arc center point. Some controllers demand how these secondary axes are located. With most, such as the Fanuc controller, the arc center point position is described as the incremental distance from the arc start point to the arc center. This method of determining arc center location is the most common. Figure 15-20 shows an example of a G02 clockwise arc. Pay close attention to the I and J values.

The center point of the arc is the distance from the start point position to the arc center position. On some other types of controls, the arc center point position is described as the absolute location of the arc center point from the workpiece zero point.

Comprehensive Programming Exercise Following is a more comprehensive example of a part program. This program will involve cutting circular interpolation. Figure 15-21 illustrates a base plate that involves linear cutting and circular cutting.

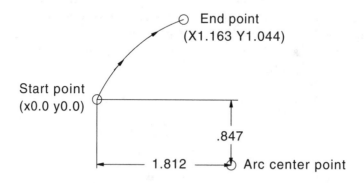

G02 x1.163 y1.044 I.812 J-.847

Figure 15–20 Sample line of code using a clockwise (G02) command. X and Y represent the end point of the arc and I and J are the incremental distance from the start point to the center of the arc.

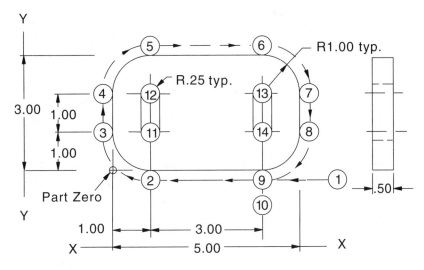

Figure 15–21 Base plate.

Following is the program for the base plate shown in Figure 15-21.

N0010 G90 G70; (Absolute programming, inch programming)

N0020 M06 T3; (Tool change, tool 3, .500 dia. endmill)

N0030 G54 X-8.500 Y-6.75; (workpiece coordinate setting)

N0040 M03 S800; (Spindle start forward, 800 RPM)

N0050 G00 X6.00 Y-.25; (Rapid traverse to position 1)

N0060 G00 Z.100; (Rapid traverse down to .100 above the part)

N0070 G01 Z-.500 F6.0; (Feed tool down to part depth)

N0080 G01 X1.00; (Feed tool to position 2)

N0090 G02 X-.25; Y1.00 I0.0 J1.25 (Clockwise arc to position 3)

N0100 G01 Y2.00; (Linear feed to position 4)

N0110 G02 X1.00 Y3.25 I1.25 J0.0; (Clockwise arc to position 5)

N0120 G01 X4.00; (Linear feed to position 6)

N0130 G02 X5.25 Y2.00 I0.0 J-1.25; (Clockwise arc to position 7)

N0140 G01 Y1.00; (Linear feed to position 8)

N0150 G02 X4.00 Y-.25 I-1.25 J0.0; (Clockwise arc to position 9)

N0160 G01 Y-1.00; (Position tool off of the part to position 10)

N0170 G28; (Return tool and all axes to home position)

N0180 M05; (Spindle off)

N0190 M06 T4; (Tool change to tool 4, .500 dia. slotting endmill)

N0200 M03 S750; (Spindle start forward, 750 revolutions per minute)

N0210 G00 X1.00 Y1.00; (Rapid traverse to position 11)

N0220 G00 Z.100; (Rapid traverse down to .100 above the part)

N0230 G01 Z-.525 F2.0; (Feed tool down through the part at 2 inches per minute feed rate)

N0240 G01 Y2.00 F5.0; (Cut the slot to position 12 at 5 inches per minute)

N0250 G01 Z.100; (Feeds the tool up to a clearance of .100 above the part)

N0260 G00 X4.00; (Rapid traverse to position 13)

N0270 G01 Z-.525 F2.0; (Feed tool down through the part at 2 inches per minute)

N0280 G01 Y1.00 F5.0; (Cut the slot to position 14 at 5 inches per minute)

N0290 G01 Z.100; (Feeds the tool up to a clearance of .100 above the part)

N0300 G28; (Return the tool and all axes to home position)

N0310 M06 T0; (Tool change to tool 0, spindle empties)

N0320 M30; (Rewind program, end program)

Tool Length Offset   Tool length offsets make it possible for the controller to adjust to different tool lengths. Every tool can be a different length, but CNC machine controls can deal with this quite easily. CNC controllers have a special memory area within the control to store tool length offsets. The tool length offset is the distance from the tool tip at home position to the workpiece z zero position (see Figure 15-22). This distance is stored in a table that the programmer can access using a G code or a tool code. A machining center that has a Fanuc control uses a G43 code. The letter address G43 code is accompanied by an H auxiliary letter and a two-digit number. The G43 tells the control to compensate the z-axis, and the H and the number tell the control which offset to call out of the tool length storage table. The tool length offset typically is accompanied by a z-axis move to activate it.

A typical tool length offset block is N0010 G43 H10 z1.0; The line number is 10. The G43 calls for a tool length offset, and the H10 is the number of the offset found in register 10 of the tool length offset file. The z1.0 will position the tool 1 inch above the workpiece zero point. It is always a good idea to have the tool length offset register number correspond to the tool number. For example, if you are using tool number 10 (T10), try to make the height offset correspond by using height offset number 10 (H10). On some types of machining center controls, such

Figure 15–22 Reason that tools need to be offset because of differences in length.

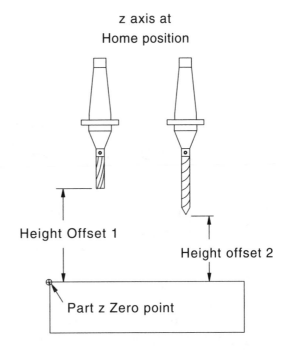

as Mazatrol, the height offset is called up with the tool number. If the program calls for tool number 10, the control automatically accesses the tool file and offsets the tool according to the tool length registered in the tool file under the tool number 10. Since machine controls vary, it is a good idea to find out how a specific control deals with variations in tool lengths.

Tool Diameter Offsets Tools also differ in diameter, tool diameter offsets compensate for this. Tool diameter offsets control the size of milled features. In the first programming example we had to compensate for the .05 diameter tool by offsetting the tool path by the radius of the tool diameter. The control can offset the path of the tool, so we can program the part just as it appears on the part print. This saves us from having to mathematically calculate the cutter path. The diameter offset also allows the programmer to program the part for any size cutter. Without the diameter offset, the programmer must state the precise size of the tool to be used and program the center of the spindle accordingly. With cutter compensation capabilities, the cutter size can be ignored and the part profile can be programmed. The exact size of the cutting tool to be used is entered into the offset file and when the offset is called, the tool path automatically is offset by the tool radius. If after the part has been inspected it is found to be too big or too small, this feature of the part can be remachined to the proper size by changing the number in the offset table. If offsets are not used, trigonometry can be used to calculate positions.

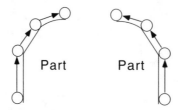

Compensation Left (G41)   Compensation Right (G42)

Figure 15–23   Tool compensation.

Tool diameter compensation allows programming the part, not the tool path. Without the aid of cutter diameter compensation, the mathematical calculations needed to program a part profile with angles and radii can be very involved. Cutter compensation can be to the right or to the left of the part profile. The offset needed is determined from a position behind the cutting tool and to the left of the programmed path or to the right.

Compensation direction is controlled by a G code (see Figure 15-23). When a compensation to the left is desired, a G41 is used. To compensate to the right, use a G42. When using the cutter compensation codes, identify the offset to be used from the offset table. A typical cutter compensation line would appear like this:

N0030 G41 D12

To use cutter compensation, the machine must make a move before actual cutting takes place (ramp on) (see Figure 15-24). This move allows the control to evaluate its present position and make the necessary adjustment from center line positioning to cutter periphery positioning. The length of the ramp on the move must be greater than the radius value of the tool. To cancel the cutter compensation and return to cutter center line programming, the programmer must make a linear move (ramp off) to invoke a cutter compensation cancellation (G40). A typical program that utilizes tool length and tool diameter compensations follows.

N0010 G70 G90; (Inch programming, absolute programming)

N0020 M06 T02; (Tool change, tool 2)

N0030 G54 X-10.250 Y-8.750 Z-7.525; (Workpiece zero setting)

N0040 M03 S800; (Spindle start clockwise, 800 revolutions per minute)

N0050 G00 X-1.00 Y-1.00; (Rapid to position 1, just off the lower left corner of the part)

N0060 G43 Z.100 H01; (Rapid down to .100 clearance above the part, invoke tool height offset)

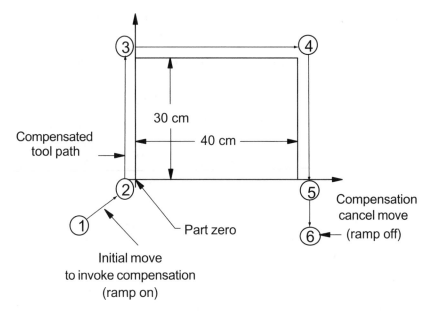

Figure 15–24 The initial move enables the machine to compensate for cutter diameter.

N0070 G01 Z-.50 F5.0; (Feed down to depth at 5 inches per minute)

N0080 G41 D2 X0.0 Y0.0; (Ramp on and invoke cutter compensation stored as D2, feed to position 2)

N0090 G01 Y3.00; (Feed to position 3)

N0100 G01 X4.00; (Feed to position 4)

N0110 G01 Y-0.25; (Feed to position 5)

N0120 G40 Y-1.00; (Ramp off and cancel cutter compensation, feed to position 6)

N0130 G00 Z0.100; (Rapid to .100 clearance above the part)

N0150 G28; (Return all axes to home position)

N0160 M06 T0; (Tool change, puts the tool away and empties the spindle)

N0170 M30 (Rewinds the program, resets the control, and ends the program)

CANNED CYCLES FOR MACHINING CENTERS

Canned cycles (fixed cycles) simplify the programming of repetitive machining operations such as drilling, tapping, and boring. Canned cycles are preprogrammed instructions that eliminate the need for many lines of programming. Programming to cut a simple drilled hole without the use of a canned cycle can take four or five lines of programming:

1. Position the x- and y-axes to the proper coordinates with a rapid traverse move (G00).
2. Position the z-axis to a clearance plane.
3. Feed the tool down to depth.
4. Rapidly position the tool back to the clearance point.

That is for one drilled hole! By using a drilling canned cycle, a hole can be cut with one line of programming. Standard canned cycles, or fixed cycles, are common to most CNC machines. See Figure 15-25 for some canned cycles for machining centers.

The most commonly used canned cycle is the G81 canned drilling cycle. It automatically performs all functions necessary to drill a hole in one line of programming (see Figure 15-26). The z-position is very important in a canned cycle because it becomes the initial z plane. The machine normally does a rapid back to the z initial position before making a rapid move to the next hole. If the tool is 12 inches above the work when the canned cycle is called, that will be the z initial plane, and the machine will rapid up to 12 inches above the work between each hole. This is a very slow program.

The canned drilling cycle consists of four moves (see Figure 15-26). Those four moves are controlled by one line of programming such as

N0010 G81 X4.500 Y2.250 Z-.375 R.100 F2.5;

G Code	Function	z-Axis	At Depth	z-Axis Return
G81	Drill	Feed	—	Rapid Traverse
G83	Peck Drill	Feed with Peck	—	Rapid Traverse
G84	Tap	Feed	Reverse Spindle	Feed
G85	Bore/Ream	Feed	Stop Spindle	Cutting Feed
G86	Bore	Feed	Stop SPindle	Rapid Traverse

Figure 15–25   Canned cycles for machining centers.

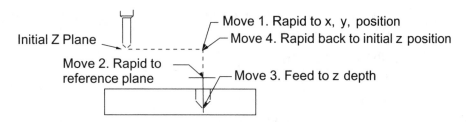

Figure 15–26 G81 canned drilling cycle.

N is the line number, G81 specifies which canned cycle, x and y are the coordinates of the hole, z is the depth of the hole, and f is the feed rate in inches per minute.

The G81 canned cycle is modal, which means that it will stay active until it is canceled by a G80. When drilling a series of holes, we only need the coordinates for the next hole to be specified. See Figure 15-27 and the following program for using a G81 canned drilling cycle.

N0010 G90 G70; (Absolute programming)

N0020 M06 T01; (Tool change, tool 1, 0.5 inch drill)

N0030 G54 X-10.500 Y-6.750 Z-16.564; (Workpiece zero setting)

N0040 M03 S1000; (Spindle start clockwise, 1000 revolutions per minute)

N0050 G00 X1.00 Y1.00; (Rapid to hole position 1)

N0060 Z1.0; (Rapid to initial level)

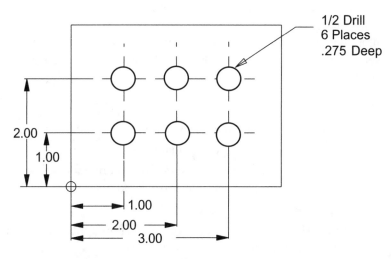

Figure 15–27 Canned drilling cycle example.

N0070 G81 Z-0.275 R0.100 F3.00; (Drill hole 1 .275 inches deep)

N0080 X2.00; (Drill hole 2)

N0090 X3.00; (Drill hole 3)

N0100 Y2.00; (Drill hole 4)

N0110 X2.00; (Drill hole 5)

N0120 X1.00; (Drill hole 6)

N0130 G80; (Cancel drilling cycle)

N0140 G28; (Return all axes to home position)

N0150 M05; (Spindle stop)

N0160 M30; (Rewind program, reset the control, and end the program)

CONVERSATIONAL PROGRAMMING

Conversational programming is a built-in feature that allows the programmer to respond to a set of questions displayed on the graphics screen. The questions guide the programmer through each phase of the machining operation. First the operator might input the material to be machined. The control uses this information to calculate speeds and feeds, but the operator can override them. Next the operator chooses the operations to be performed and inputs the geometry. Operations such as pocket milling, grooving, drilling, and tapping can be performed.

With each response to the questions, more questions are presented until the operation is complete. By answering the questions, the programmer is filling in the variables in a canned or preprogrammed cycle. This is quicker than methods using machine code language. There is no standard conversational part programming language, and each system can be quite different. However, once the programmer completes the conversational program, many CNC controls convert the conversational language into standard EIA/ISO machine language.

QUESTIONS

1. Describe computer numerical control.
2. Name the two basic types of positioning.
3. Draw a picture of the two most common computer numerical control machines and label the basic axes associated with each.
4. Define the zero or origin point.
5. Draw a diagram of a closed-loop axis. Thoroughly explain the velocity and position loops.
6. Thoroughly describe how a servo is homed.

7. Describe the incremental positioning mode and give an example.
8. Describe the absolute positioning mode and give an example.
9. For what are M codes used?
10. For what are G codes used?
11. What does *modal* mean?
12. Program the following part below. T1 is a .25 drill. Tool 2 is a .5 end-mill. Write a program to mill the outside of the part, mill the slot, and drill the holes.

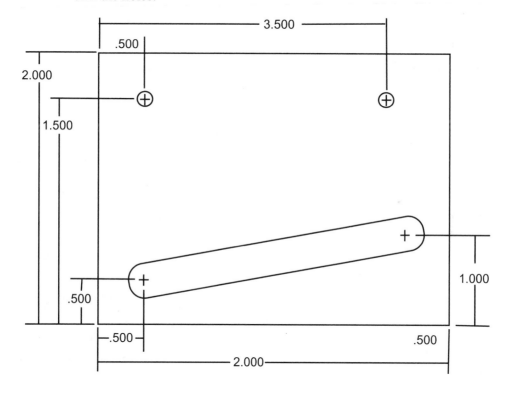

Installation, Maintenance, and Troubleshooting

16

Proper installation of automated systems is crucial. The safety of people and machines is at stake. Troubleshooting and maintenance of systems becomes more crucial as systems become more automatic, more complex, and more expensive. An enterprise cannot afford to have a system down for any length of time. The technician must be able to find and correct problems quickly.

OBJECTIVES

Upon completion of this chapter you will be able to:
1. Describe crucial safety considerations in troubleshooting and maintaining systems.
2. Explain the terms *noise, snubbing, suppression,* and *single-point ground.*
3. Explain correct installation techniques and considerations.
4. Explain proper grounding techniques.
5. Explain noise reduction techniques.
6. Explain a typical troubleshooting process.

INSTALLATION, MAINTENANCE, AND TROUBLESHOOTING

Installation, maintenance and troubleshooting of an automated system are among the most crucial phases of any project. It is the point at which the hopes and fears of the engineers and technicians are realized. It can be a frustrating, exciting, and rewarding time. Installation, maintenance, and troubleshooting of any automated system depends highly on the quality of the associated documentation, which unfortunately is often an afterthought. Documentation should be developed accu-

371

rately and completely as the system is developed. Downtime can be significantly reduced with good documentation. Documentation should include the following:

Description of the overall system.

Block diagram of the entire system.

Program list including cross-referencing and clearly labeled I/O.

Printout of the PLC or control device's memory showing I/O and variable usage.

Complete wiring diagram.

Description of peripheral devices and their manuals.

Operator's manual including startup and shutdown procedures; notes concerning past maintenance.

INSTALLATION

Proper installation of a PLC is crucial. It must be wired so that the system is safe for the workers and so that the devices are protected from overcurrent situations. Proper fusing within the system is important. The PLC must also be protected from dust, coolant, chips, and other contaminants in the air of the application's environment. The proper choice of enclosures can protect the PLC. Figure 16-1 is a block diagram of a typical installation.

Enclosures

PLCs are typically mounted in protective cabinets. The National Electrical Manufacturers Association (NEMA) has developed standards for enclosures (see

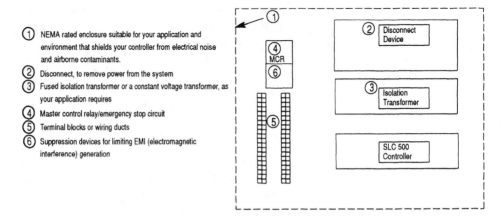

① NEMA rated enclosure suitable for your application and environment that shields your controller from electrical noise and airborne contaminants.

② Disconnect, to remove power from the system

③ Fused isolation transformer or a constant voltage transformer, as your application requires

④ Master control relay/emergency stop circuit

⑤ Terminal blocks or wiring ducts

⑥ Suppression devices for limiting EMI (electromagnetic interference) generation

Figure 16–1   Block diagram of a typical PLC control cabinet. (*Courtesy Rockwell Automation.*)

Protection Against	Enclosure Type												
	1	2	3	3-r	3-s	4	4-x	5	6	6-p	1-1	1-2	1-3
Accidental contact with enclosed equipment	*	*	*	*	*	*	*	*	*	*	*	*	*
Falling dirt	*	*				*	*	*	*	*	*	*	*
Falling liquids, light splashing		*				*	*		*	*	*	*	*
Dust, lint, fibers, (noncombustible, nonignitable)						*	*	*	*	*		*	*
Windblown dust			*		*	*	*		*	*			
Hose down and splashing water						*	*		*	*			
Oil and coolant seepage												*	*
Oil or coolant spraying or splashing													*
Corrosive agents							*			*	*		
Occasional temporary submersion									*	*			
Occasional prolonged submersion										*			

Figure 16–2   NEMA standards for cabinet enclosures.

Figure 16-2). *Enclosures are used to protect the control devices from the environment of the application.* A cabinet type is chosen based on the characteristics of the environment. Cabinets typically protect the PLC from airborne contamination, and metal cabinets protect from electrical noise.

The PLC and other devices mounted in the cabinet must be able to perform at the temperatures required. Remember that the temperature in the cabinet will be higher than in the application because of the heat generated by the devices in the cabinet. In this case, blower fans should be placed inside the enclosure to increase air circulation and reduce hot spots. These fans should filter the incoming air to prevent contaminants from entering the cabinet.

The main consideration is that space around the PLC in the enclosure is adequate. This will allow air to flow around the PLC. The PLC manufacturer provides installation requirements in the hardware manual. Figure 16-3 is an example of the proper mounting for an AB SLC 500. Note that the dimensions shown from each side of the PLC to the cabinet are *minimum* distances. Check the temperature specifications for the PLC. In most cases fans are unnecessary. Proper clearances around devices is normally sufficient for heat dissipation.

Care must be taken when drilling holes in the cabinet for mounting components so that chips do not fall into the PLC or other components. Metal chips or wire clippings could cause equipment to short circuit or could cause intermittent or permanent problems.

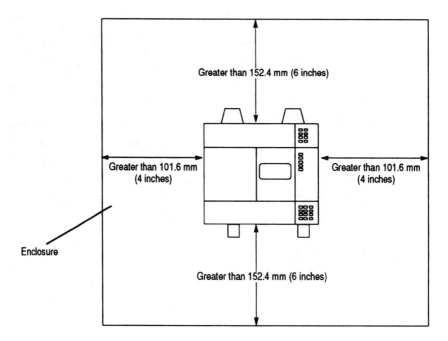

Greater than 152.4 mm (6 inches)

Greater than 101.6 mm
(4 inches)

Greater than 101.6 mm
(4 inches)

Enclosure

Greater than 152.4 mm (6 inches)

Figure 16–3 Proper mounting for an AB SLC 500. (*Courtesy Rockwell Automation.*)

Wiring

Proper wiring of a system involves choosing the appropriate devices and fuses (see Figure 16-4). Normally, three-phase power (typically 480 volts) are used in manufacturing and is the electrical supply for the control cabinet.

The three-phase power is connected to the cabinet via a mechanical disconnect that is turned on or off by the use of a lever on the outside of the cabinet. This disconnect should be equipped with a lockout (see the earlier chapter on lockout/tagout) to prevent anyone from accidentally applying power during operation. This three-phase power must be fused, a fusible disconnect normally is used. This means that the mechanical disconnect has fusing built in to ensure that too much current cannot be drawn. Figure 16-4 shows the fuses installed after the disconnect.

The three-phase power is then connected to a contactor that turns all power to the control logic off in an emergency. The contactor also called a *master control relay* is attached to hard-wired emergency circuits in the system. If someone hits an emergency stop switch, the contactor drops out and will not supply power. The hard-wired safety should always be used in systems.

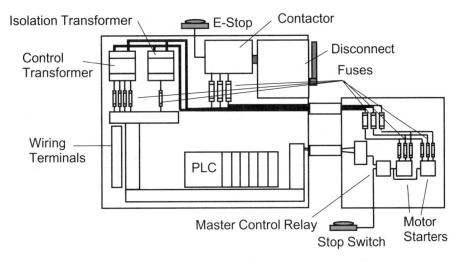

Figure 16–4 Block diagram of a typical control cabinet wiring scheme.

Master Control Relay

A hard-wired master control relay (MCR) is used for emergency control shutdown (see Figure 16-5). The MCR must be able to inhibit all machine motion by removing power to the machine's I/O devices when the MCR relay is deenergized.

If a DC power supply is used, the power should be interrupted on the load side rather than on the AC supply side. This provides faster power shutdown. The DC power supply should be powered directly from the fused secondary of the transformer. DC power from the power supply is then connected to the devices through a set of MCR contacts.

Emergency stop switches can be placed at multiple locations to provide full safety for anyone in the area. Switches may include limit-type switches for overtravel conditions and mushroom-type push-button switches. They are wired in series so that if any one of the switches is activated, the MCR is deenergized, removing power from all input and output circuits.

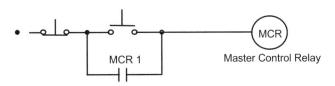

Figure 16–5 Master control relay circuit.

Severe injury and/or damage can occur if these circuits are altered or bypassed to defeat their function. Switches should be tested periodically to ensure they will still stop all machine motion if used.

The main power disconnect switch should be placed outside the cabinet in a convenient and easily accessible place for operators and maintenance personnel. The MCR is not a substitute for a disconnect to the PLC but is used to quickly deenergize I/O devices. Note that the system in Figure 16-6 provides for a mechanical disconnect for output module power and a hard-wired stop switch for disconnecting the system power. The programmer should also provide for an orderly shutdown in the PLC control program.

The three-phase power must then be converted to single-phase for the control logic. Power lines from the fusing are connected to transformers. In Figure 16-7 there are two transformers: an isolation transformer and a control transformer. The isolation transformer cleans up the power supply for the PLC. It is normally used when high-frequency conducted noise occurs in or around power distribution equipment. Isolation transformers are also used to step down the line voltage.

The control transformer supplies other control devices in the cabinet. The lines from the power supplies are then fused to protect the devices they will supply. These individual device circuits should be provided with their own fuses to match the current draw. DC power is usually required. If necessary a small power supply is typically used to convert the AC to DC. Motor starters are typically mounted in separate enclosures to protect the control logic from the noise these devices generate.

Within the cabinet certain wiring conventions are typically used: red for control wiring, black wiring for three-phase power, blue for DC, and yellow for showing that the voltage source is separately derived power (outside the cabinet).

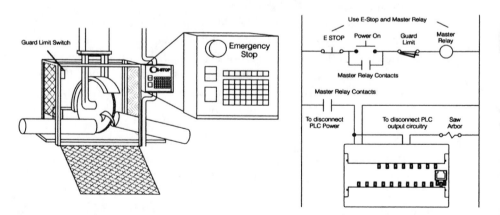

Figure 16–6   Hardwired emergency stop (E-stop) and master control relay. (*Courtesy PLC Direct by Koyo.*)

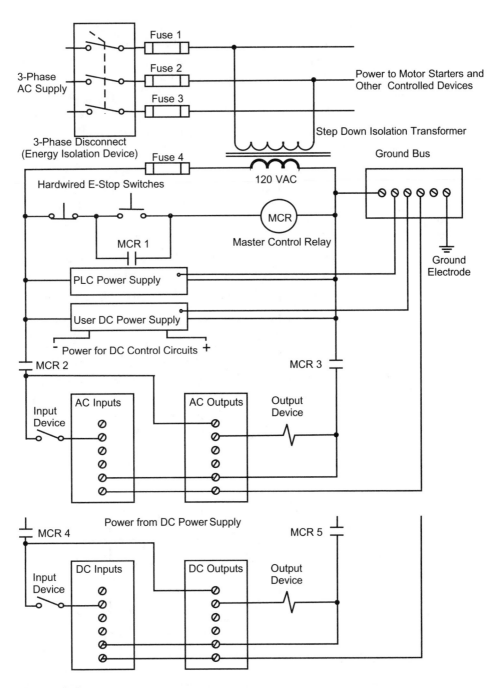

Figure 16–7 Control wiring diagram.

Signal wiring should be run separately from 120 volt wiring. Signal wiring is typically low voltage or low current and could be affected by being too close to high-voltage wiring. When possible, the signal wires should run in separate conduit. Some conduit is internally divided by barriers to isolate signal wiring from higher voltage wiring.

Wiring terminals supplies voltage to the PLC. The user can often configure the PLC to accept different voltages (see Figure 16-8). The PLC in this figure can be configured to accept either 220 or 110 volts.

Wiring Guidelines   The following guidelines that Siemens Industrial Automation, Inc. suggests should be considered when wiring a system:

Always use the shortest possible cable.

Use a single length of cable between devices. Do not connect pieces of cable to make a longer cable. Avoid sharp bends in wiring.

Avoid placing system and field wiring close to high-energy wiring.

Physically separate field input wiring, output wiring, and other types of wiring.

Separate DC and AC wiring when possible.

A good ground must exist for all components in the system (0.1 ohm or less).

If long return lines to the power supply are needed, do not use the same wire for input and output modules. Separate return lines will minimize the voltage drop on the return lines of the input connections.

Use cable trays for wiring.

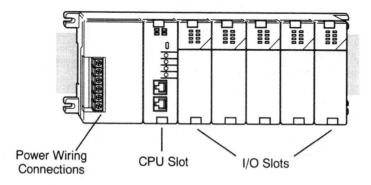

Power Wiring Connections   CPU Slot   I/O Slots

Figure 16–8   Diagram of PLC power supply wiring, CPU, and I/O. (*Courtesy PLC Direct by Koyo.*)

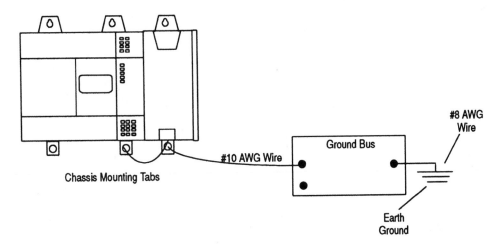

Figure 16–9 Grounding of an SLC 500 controller with a two-slot expansion chassis. (*Courtesy Rockwell Automation.*)

Grounding

Proper grounding is essential for the safe and proper operation of a system. It also limits the effects of noise due to electromagnetic interference (EMI). The appropriate electrical codes and ordinances must be checked to ensure compliance with minimum wire sizes, color coding, and general safety practices.

Connect the PLC and components to the subpanel ground bus (see Figure 16-9). Ground connections should run from the PLC chassis and the power supply for each PLC expansion unit to the ground bus. The connection should exhibit very low resistance. The subpanel ground bus should be connected to a single-point ground, such as a copper bus bar, or a good earth ground reference. Impedance between each device and the single-point termination must be low. A rule of thumb is less than 0.1 ohm DC resistance between the device and the single-point ground. This can be accomplished by removing the anodized finish and using copper lugs and star washers.

The PLC manufacturer provides details on installation in the hardware installation manual for its equipment. The National Electrical Code is an authoritative source for grounding requirements.

Grounding Guidelines See Figure 16-10 for examples of ground connections. Grounding braid and green wires should be terminated at both ends with copper lugs to provide good continuity. Lugs should be crimped and soldered. Copper

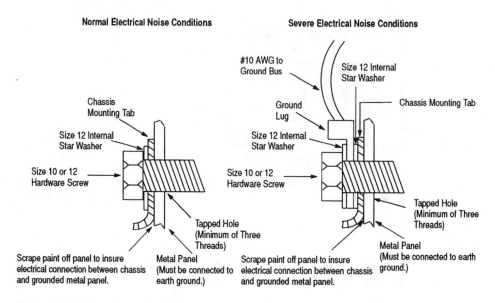

Figure 16–10   Ground connections. (*Courtesy Rockwell Automation.*)

No. 10 or 12 bolts should be used for fasteners that provide an electrical connection to the single-point ground, to device mounting bolts, and braid termination bolts for subpanel and user-supplied single-point grounds. Tapped holes should be used rather than nuts and bolts. A minimum number of threads is required for a solid connection.

Paint, coatings, or corrosion must be removed from the areas of contact by external toothed lock washers (star washers). This practice should be used for all terminations: lug to subpanel, device to lug, device to subpanel, subpanel to conduit, and so on. Ensure that the appropriate electrical codes and ordinances comply with minimum wire sizes, color coding, and general safety practices.

Electrical Noise

Electrical noise is unwanted electrical interference that affects control equipment. Control devices used today utilize microprocessors to constantly fetch data and instructions from memory. Noise can cause the microprocessor to misinterpret an instruction or fetch bad data, resulting in minor problems to severe damage to equipment and people.

Noise is caused by a wide variety of manufacturing devices. Those that switch high voltage and current are the primary sources of noise. These include large motors and starters, welding equipment, contactors that turn devices on and off, power line disturbances, and transmitted noise or ground loops. Power line disturbances are generally caused by devices that have coils including relays, contactors, starters, clutches/brakes, and solenoids. When they are switched off, the devices create a line disturbance. Power line disturbances can normally be overcome through the use of line filters. Surge suppressors such as MOVs or an RC network across the coil can limit the noise.

Transmitted noise caused by devices that create radio frequency noise is generally caused in high current applications such as welding. When contacts that carry high current are open, they generate transmitted noise that can disrupt application wiring carrying signals. For example, in severe cases, false signals can be generated on the signal wiring. This problem can often be overcome by using twisted-pair shielded wiring and connecting the shield to ground.

Transmitted noise can also "leak" into control cabinets through holes for switches and wiring. The effect can be reduced by properly grounding the cabinet.

Ground loops are the noise problems that are often difficult to find. Quite often they are intermittent problems. They generally occur when multiple grounds exist. The farther the grounds are apart, the more likely the problem. The power supply earth and the remote earth can create unpredictable results, especially in communications.

High voltage such as that resulting from turning off an inductive load can cause trouble for the PLC. Lack of surge suppression can contribute to processor faults and intermittent problems. Noise can also corrupt RAM memory and may cause intermittent problems with I/O modules. Many of these problems are difficult to identify and/or repair because of their sporadic nature. Excessive noise can also significantly reduce the life of relay contacts. The two main ways to deal with noise are suppression and isolation.

Noise Suppression   Suppression attempts to deal with a device that is generating noise. A suppression network can be installed to limit voltage spikes. Surge suppression circuits connect directly across the load device to reduce the arcing of output contacts. Some PLC modules include protection circuitry to protect against inductive spikes.

Both AC and DC loads can be protected against surges (see Figure 16-11). A diode is sufficient for DC load devices. A 1N4004 can be used in most applications. Surge suppressors can also be used. Noise suppression called *snubbing* suppresses the arcing of mechanical contacts caused by turning inductive loads off (see Figure 16-12). Inductive devices include relays, solenoids, motor starters, and motors. Surge suppression should be used on all coils.

An RC or a varistor circuit can be used across an inductive load to suppress noise (1000 ohm, 0.2 microfarad). These components must be sized appropriately to meet

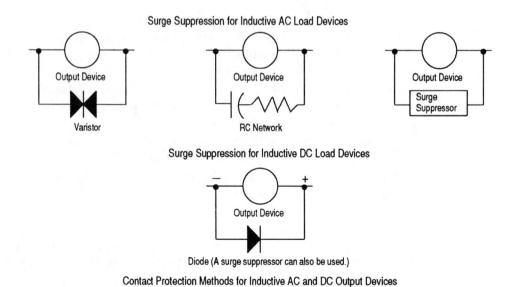

Figure 16–11   Surge suppression methods for AC and DC loads. (*Courtesy Rockwell Automation.*)

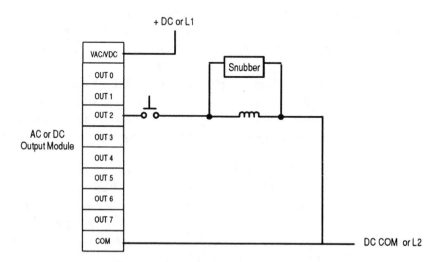

Figure 16–12   Snubbing. (*Courtesy Rockwell Automation.*)

the characteristic of the inductive output device being used. The installation manual for a specific PLC has information concerning proper noise suppression.

Noise Isolation The other way to deal with noise is isolation, which physically separates the device or devices that cause trouble from the control system. The enclosure can help to do this. In many cases field wiring must be placed in very noisy environments to allow sensors to monitor the process. This presents a problem especially with low voltages. Shielded twisted-pair wiring should be used for the control wiring in these cases. The shielding should be grounded at only one end. The shield should be grounded at the single-point ground.

To reduce the effects of noise, follow these steps.

Properly mount the PLC within a suitable enclosure.

Properly route all wiring.

Properly ground all equipment.

Install suppression to noise-generating devices.

INDUSTRIAL CONTROLLER MAINTENANCE

Industrial controllers are designed to be very reliable. They are used in automated systems that are very costly to operate, making downtime very costly in industry. The technician will be expected to keep the systems in operation and downtime to a minimum. Industrial controllers have been designed to provide low maintenance, but some tactics help reduce downtime.

It is important to keep controllers relatively clean. They are normally mounted in air-cooled enclosures. A fan in the wall of the enclosure circulates fresh air through the cabinet to cool the components. This can lead to the accumulation of dust, dirt and other contaminants. The fans for such enclosures must have adequate filters that are cleaned regularly. A preventive maintenance schedule should include a check of the enclosure for the presence of contaminants and for loose wires or termination screws that could cause problems later. Modules should also be checked to ensure that they are securely seated in the backplane, especially in high-vibration environments.

Many control devices have battery backup. Long-life lithium batteries with lifetimes of two or more years are usually used, but they will fail eventually. Therefore, preventive maintenance schedules should include checking batteries and purchasing replacements. Replacing batteries once a year is an inexpensive investment to avoid costly downtime and potentially larger problems. Preventive maintenance also includes keeping an inventory of spare parts to minimize downtime.

TROUBLESHOOTING

The first consideration in troubleshooting and maintaining systems is safety. Approximately less than one-third of all system failures are caused by the PLC. Input and output devices are responsible for approximately 50 percent of failures.

The technician must always be aware of the possible outcomes of changes. A few years ago, a technician was killed when he isolated the problem to a defective sensor. He bypassed the sensor, and the system restarted with him in it. He fixed the system but at a very high price.

Troubleshooting is a relatively straightforward process in automated systems. The first step is to think. This may seem rather basic, but many people jump to improper, premature conclusions and waste time finding problems. Examine the problem logically. Think the problem through using common sense. Troubleshooting is much like the game 20 Questions; every question should help isolate the problem. In fact, every question should eliminate about half of the potential causes.

Next use the available resources to check the reason suspected (theory) as the source of the problem. Often the error-checking on the PLC modules is sufficient. The LEDs on PLCs, CPUs, and modules can provide immediate feedback on what is happening. The LEDs can indicate blown fuses and many other problems.

Consider the usual causes when an output is not turning on when it should: defective parts in the output device, the PLC, the input, or wiring; and faulty ladder logic. The module I/O LEDs provide the best source of answers (see Figures 16-13 and 16-14).

The next step is to isolate the problem further. A multimeter is invaluable at this point. If the PLC output is off, a meter reading should show the full voltage

Output Condition	Output LED	Ladder Status	Probable Problem
On	On	True	None
Off	Off	True	Bad Fuse or Bad Output Module
Off	Off	False	None
Off	On	True	Wiring to Output Device or Bad Output Device

Figure 16–13   Isolation of a PLC problem by comparing the states of inputs/outputs, indicators, and the ladder status.

Figure 16–14 An output
tested to isolate the problem.

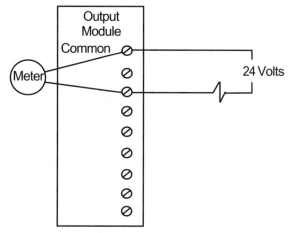

with which the device is turned on. If the output is used to supply 115 volts to a motor starter, the meter should show the full 115 volts between the output terminal and the common on the PLC module (see Figure 16-14).

If the PLC output is on, the meter should read zero volts because the output acts as a switch. The voltage across a switch should read zero volts. If the switch is open, it should read the full voltage. If the reading indicates no voltage in either case, the wiring and power supply should be checked. If the wiring and power

Actual Input Condition	Input Module LED Status	Ladder Status		Probable Problem
Off	Off	False	True	None
Off	On	True	False	Short in the Input Device or Wiring or a Bad Input Module
On	Off	False	True	Wiring/Power to the I/O Module
On	On	False	True	I/O Module
On	On	True	False	None

Figure 16–15 Chart for troubleshooting input.

supply are good, the device is the problem. Depending on the device, fuses or overload protection may be present.

Suppose that the output LED does not turn on. The input side should be checked. If the input LED is on, the sensor or other input device is operational (see Figures 16-15 and 16-16). Test whether the PLC's CPU really sees the input as true. At this point, a monitor such as a hand-held programmer or a computer is required. The ladder should be monitored under operation. (*Note:* Many PLCs allow the outputs to be disabled for troubleshooting. This is the safe way to proceed.) Check the ladder to see whether the contact is closing. If it is seeing the input as false, the problem is probably a defective PLC module input. See Figures 16-17 and 16-18 for troubleshooting input and output modules.

Potential Ladder Diagram Problems

Some PLC manufacturers allow output coils to be used more than once in ladders. This means that multiple conditions can control the same output coil. The author has seen technicians who were sure that they had a defective output module because when they monitored the ladder, the output coil was on but the actual output LED was off. The technician had inadvertently programmed the same output coil twice with different input logic. The rung he was monitoring was true, but one down in the ladder was false, so the PLC kept the output off.

Another potential problem with ladder diagramming is that the problem can be intermittent. Timing is crucial in ladder diagramming. Devices are often exchanging signals to signify that an action has occurred. For example, a robot finishes its cycle and sends a digital signal to the PLC to let it know. The robot's programmer must be

Figure 16–16 Isolation of input problems.

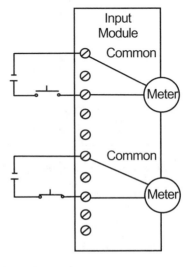

Input LED State	Real-World State of Device	Condition	Cause	Action
Off	Off	Program operates as if the input is on.	Input is forced on in the program.	Check the forced I/O on the processor and remove forces, verify wiring. Try another input or replace the module.
			Input circuit is damaged.	Verify wiring. Try another input or replace the module.
	Off	Input device will not turn on.	The input device is damaged.	Verify device operation or replace if defective.
On	On	Input device will not turn off.	Device is damaged or shorted.	Verify device operation or replace if defective.
		Program operates as if the input device is off.	Input is forced off in program.	Check the forced I/O on the processor and remove forces.
			Damaged input circuit.	Verify the wiring. Try another input or replace the module.
	Off	Program operates as though the input is on and/or the input circuit will not turn off.	Damaged input circuit.	Verify the wiring. Try another input or replace the module.
			Input device is damaged or shorted.	Verify device operation or replace if defective.
			The leakage current of the input device exceeds the input circuit specification.	Use load resistor to bleed-off current.

Figure 16–17   Input troubleshooting.

aware of the timing considerations. If the programmer turns it on for only one step, the output may be on for only a few milliseconds. The PLC may be able to "catch" the signal every time, or it may occasionally miss it because of the length of the scan time of the ladder diagram. This can be a difficult problem to find because it happens only occasionally. The better way to program is to handshake instead of relying on the PLC seeing a periodic input. The programmer should have the robot output stay on until the PLC acknowledges it. The robot turns on the output and waits for an input from the PLC to ensure that the PLC saw the output from the robot.

Output LED State	Real-World State of Device	Condition	Cause	Action
Off	On	Output device will not turn off but the program indicates it is off.	Incorrect wiring.	Check the wiring. Disconnect the device and test.
			Output device is damaged or shorted.	Verify device, replace if necessary.
			Output circuit is defective.	Check the wiring. Try another output circuit, replace module if possible.
	Off	Program shows that the output circuit is on or the output circuit will not turn on.	Damaged output circuit.	Use the force function to turn the output on. If the output turns on, there is a programming problem. If not, there is an output circuit problem. Try a different output circuit and replace the module if necessary.
			Programming problem.	Check for duplicate output addresses. If subroutines are being used, outputs are in their last state when not executing subroutines. Use the force function to force the output on. If the output does not turn on, the output circuit is damaged. Try a different output circuit and replace the module if necessary. If the output does force on, check for a programming problem.
			The output is forced off in the program.	Check the processor forced I/O or force LED and remove forces.
On	On	The program shows the output circuit off or the output circuit will not turn off.	Programming problem.	Check for duplicate output addresses. If subroutines are being used, outputs are in their last state when not executing subroutines. Use the force function to force the output off. If the output does not turn off, the output circuit is damaged. Try a different output circuit and replace the module if necessary. If the output does force off, check for a logic or programming problem.
			Damaged output circuit.	Use the force function to turn the output off. If the output turns off, there is a programming problem. If not, there is an output circuit problem. Try a different output circuit and replace the module if necessary.
			Output is forced on in the program.	Check the processor forced I/O or force LED and remove forces.
	Off	Output device will not turn on but the program indicates it is on.	Incorrect wiring or an open circuit.	Check the wiring and the common connections.
			Defective output circuit.	Try a different output circuit and replace the module if necessary.
			Low or no voltage across the load.	Measure the source voltage.
			Output device incompatibility.	Check the specifications—sink/source etc.

Figure 16–18 Output troubleshooting.

388

QUESTIONS

1. What is a NEMA enclosure?

2. Why should enclosures be used?

3. Describe how to choose an enclosure.

4. What is a fusible disconnect?

5. What is a contactor?

6. What is the purpose of an isolation transformer?

7. What is the major cause of failure in systems?

8. Describe a logical process for troubleshooting.

9. Describe proper wire grounding techniques.

10. Draw a block diagram of a typical control cabinet.

11. Describe at least three precautions that should be taken to help reduce the problem of noise in a control system.

12. A technician has been asked to troubleshoot a system. The output device is not turning on for some reason. The output LED is working as it should. The technician turns the output on and places a meter over the PLC output that reads 115 volts. The device is not running. What is the most likely problem? How might it be fixed?

13. A technician has been asked to troubleshoot a system. The PLC does not seem to be receiving an input because the output it controls is not turning on. The technician notices that the input LED is not turning on and that the output indicator LED on the sensor seems to be working. A meter placed across the input with the sensor on 24 volts, senses, but the input LED is off. What is the most likely problem?

14. A technician is asked to troubleshoot a system. An input device seems to be defective. The technician notices that the input LED is never on. A meter placed across the PLC input reads zero volts. The technician removes and tests the sensor. (The LED on the sensor comes on when the sensor is activated.) The sensor is fine. What is the most likely problem?

15. A technician is asked to troubleshoot a system. An output device is not working. The technician notices that the output LED seems to be working fine. A meter is placed over the PLC output with the output on reads zero volts. Describe what the technician should do to find the problem.

16. What items should be included in system documentation?

Glossary

Absolute pressure: The pressure of a gas or liquid measured in relation to a vacuum (zero pressure).

Acidity: A measure of the hydrogen ion content of a solution.

AC input module: A module that converts a real-world AC input signal to the logic level required by the PLC processor.

AC output module: A module that converts the processor logic level to an AC output signal to control a real-world device.

Accumulated value: The present count or time of timers and counters.

Accuracy: The deviation between the actual position and the theoretical position.

Action: Defines what is done to regulate the final control element to effect control of a controller action. Types include on-off, proportional, integral, and derivative.

Actuator: Output device such as an air valve or cylinder normally connected to an output module.

Address: A number used to specify a storage location in memory.

Alkalinity: A measure of the hydroxyl ion content of a solution.

Ambient temperature: The natural temperature in the environment such as a PLC in a cabinet near a steel furnace.

Analog: A signal with a smooth range of possible values. For example, a temperature that could vary between 60 and 300 degrees would be analog in nature.

ANSI: American National Standards Institute.

ASCII: American Standard Code for Information Interchange. A coding system that represents letters and characters. Seven-bit ASCII can represent 128 different combinations, and eight-bit ASCII (extended ASCII) can represent 256 different combinations.

Asynchronous communications: A communication method that uses a series of bits including a start bit, data bits (7 or 8), a parity bit (odd, even none, mark, or space), and stop bits (1, 1.5, or 2) to send data between devices. One character is transmitted at a time. RS-232 is the most common.

Atmospheric pressure: The pressure exerted on a body by the air. It is equal to 14.7 pounds per square inch at sea level.

Backplane: A bus is a printed circuit board with sockets that accept various modules in the back of a PLC chassis.

Batch process: A process that processes a given amount of material in one operation to produce what is required.

Baud rate: The speed of serial communications, the number of bits per second transmitted. For example, RS-232 is normally used with a baud rate of 9600 (about 9600 bits per second). It takes about 10 bits in serial to send an ASCII character so that a baud rate of 9600 transmits about 960 characters per second.

Bellows: A pressure-sensing element consisting of a metal cylinder that is closed at one end. The difference in pressure between the inside and outside of the cylinder causes the cylinder to expand or contract its length.

Beta ratio: The diameter of an orifice divided by the internal diameter of the pipe.

BEUG (BITBUS European Users Group): A nonprofit organization devoted to spreading the BITBUS technology and organizing a basic platform where people using BITBUS can share application experiences.

Binary: A base two number system using ones and zeros to represent numbers.

Binary-coded decimal (BCD): A number system in which each decimal number is represented by four binary bits. For example, the decimal number 967 is represented by 1001 0110 0111 in BCD.

Bit: A binary digit; the smallest element of binary data; either a 0 or a 1.

BITBUS: BITBUS was created by Intel in 1983 and promoted as a standard (IEEE-1118 1990) in 1990 by a special committee of the IEEE. It is one of the most widely used fieldbuses.

Boolean: A logic system that uses operators such as AND, OR, NOR, and NAND utilized by PLCs, although it is usually made invisible by the programming software for the ease of the programmer.

Bounce: An undesirable effect that is the erratic make and break of electrical contacts.

Branch: A parallel logic path in a ladder diagram.

Byte: Eight bits or two nibbles. (A nibble is four bits.)

Calibration: A procedure used to determine, correct, or check the absolute values corresponding to the graduations on a measuring instrument.

Cascade: A programming technique used to extend the range of timers and counters.

Celsius: The centigrade temperature scale on which the freezing point of water is 0 degrees and the boiling point is 100 degrees.

CENELEC: The European Committee for Electrotechnical Standardization develops standards that apply dimensional and operating characteristics of control components.

Central processing unit (CPU): The microprocessor portion of the PLC that handles the logic.

Closed loop: A system that has feedback and can automatically correct for errors.

Cold junction: The point at which a pair of thermocouple wires is held at a fixed temperature; also called the *reference junction*.

Color mark sensor: A sensor designed to differentiate between two different colors on the basis of contrast between the two colors.

Complement: The inverse of a digital signal.

CMOS (complementary metal-oxide semiconductor): The integrated circuits that consume very little power and have good noise immunity.

Compare instruction: PLC instruction used to test numerical values for equal, greater than, or less than relationships.

Contact: A symbol used in programming PLCs to represent inputs; can be normally open and normally closed; the conductors in electrical devices such as starters.

Contactor: A special-purpose relay used to control large electrical current.

CSA (Canadian Standards Organization): The organization that develops standards, tests products, and provides certification for a wide variety of products.

Current sinking: An output device (typically an NPN transistor) that allows current flow from the load through the output to ground.

Current sourcing: An output device (typically a PNP transistor) that allows current flow from the output through the load and then to ground.

Cyclic Redundancy Check (CRC): A calculated value based on the content of a communication frame. It is inserted in the frame to check data accuracy after receiving the frame across a network. BITBUS uses the standard SDLC CRC.

Dark-on: A photosensor's sensor output when no object is sensed.

Data highway: A communications network for devices such as PLCs. They are normally proprietary, which means that only like devices of the same brand can communicate over the highway. Allen Bradley calls its PLC communication network Data Highway.

Data table: A consecutive group of user references (data) of the same size that can be accessed with table read/write functions.

Debugging: The process of finding problems (bugs) in any system.

Derivative gain: The derivative in a PID system that acts based on the rate of change in the error. It has a damping effect on the proportional gain.

DeviceNet: An industrial bus for field-level devices; based on the Controller Area Network (CAN) chip.

Diagnostics: The software routines that aid in identifying and finding problems and fault conditions in a system.

Digital output: An output that can have two states: on or off; also called *discrete outputs*.

Distributed processing: A concept that allows individual discrete devices to control their area and still communicate to the others via a network; its control takes the processing load off the "host" system.

Documentation: The descriptive paperwork that explains a system or program so that the technician can understand or change the system, install devices on it, troubleshoot it, and maintain it.

Downtime: The time a system is not available for production or operation because of breakdowns in systems.

EEPROM: Electrically erasable programmable read-only memory.

Energize: The instruction that causes a bit to be a 1. This turns an output on.

Examine-off: The (normally closed) contact used in ladder logic; is true (or closed) if the real-world input associated with it is off.

Examine-on: The (normally open) contact used in ladder logic programming; is true (or closed) if the real-world input associated with it is on.

Expansion rack: A rack added to a PLC system when the application requires more modules than the main rack can contain; sometimes used to permit an I/O to be remotely located from the main rack.

False: The disabled logic state (off).

Fault: A failure in a system that prevents its normal operation.

Firmware: The combination of software and hardware that is a series of instructions contained in read-only memory (ROM) used for the operating system functions.

Flowchart: An illustration used to make program design easier.

Force: The change in the state of actual I/O by changing the bit status in the PLC; normally used to troubleshoot a system.

Frame: A packet of bits to be transmitted across a network; contains a header, user data, and an end of frame; must contain all necessary information to enable the sender and receiver(s) of the communication to decode the user's data and to ensure that the data are correct.

Full duplex: Communication scheme in which data flows in both directions simultaneously.

Ground: Direct connection between equipment (chassis) and earth ground.

Half duplex: Communication scheme in which data can flow in both directions but in only one direction at a time.

Hard contacts: A physical switch connection.

Hard copy: A printed copy of computer information.

Hexadecimal: A numbering system that utilizes base 16.

Host computer: The computer to which devices communicate; may download or upload programs or be used to program the device.

Hysteresis: A dead band purposely introduced to eliminate false reads in a sensor. An encoder hysteresis is introduced in the electronics to prevent ambiguities if the system happens to dither on a transition.

IEC (International Electrotechnical Commission): An organization that develops and distributes recommended safety and performance standards.

IEEE: The Institute of Electrical and Electronic Engineers.

Image table: An area used to store the status of input and output bits.

Incremental: A description of an encoder that provides logic states of 0 and 1 for each successive cycle of resolution.

Instruction set: The instructions available to program the PLC.

Intelligent I/O: The PLC modules that have a microprocessor built in such as a module that controls closed-loop positioning.

Interfacing: The act of connecting a PLC to external devices.

Integral gain: The device in a PID system that corrects for small errors over time; eliminates offset and permanent error.

I/O (input/output): A device used to speak about the number of inputs and outputs needed for a system or that a particular programmable logic controller can handle.

IP rating: The rating system established by the IEC that defines the protection offered by electrical enclosures; similar to the NEMA rating system.

Isolation: A process that segregates real-world inputs and outputs from the central processing unit to ensure the CPU's protection even if a major problem with real-world inputs or outputs (such as a short) occurs, normally provided by optical isolation.

K: The abbreviation for the number 1000; in computer language equal to 2^{10}, or 1024.

Keying: The technique to ensure that modules are not put in the wrong slots of a PLC.

Ladder diagram: The programmable controller language that uses contacts and coils to define a control sequence.

LAN: *See* Local area network.

Latch: An instruction used in ladder diagram programming to represent an element that retains its state during controlled toggle and power outage.

Leakage current: A small amount of current that flows through load-powered sensors.

LED (light-emitting diode): A solid-state semiconductor that emits red, green, or yellow light or invisible infrared radiation.

Light-on sensor: A photosensor's output that is on when an object is sensed.

Linear output: The analog output.

Line driver: A differential output driver intended for use with a differential receiver; usually used when long lines and high frequency are required and noise may be a problem.

Line-powered sensor: A device powered from the power supply. These are normally 3-wire sensors although 4-wire models also exist. The third wire is used for the output signal.

Load: A device through which current flows and produces a voltage drop.

Load-powered sensor: A sensor with two wires. A small leakage current flows through the sensor even when the output is off. The current is required to operate the sensor electronics.

Load resistor: A resistor connected in parallel with a high-impedance load to enable the output circuit to output enough current to ensure proper operation.

Local area network (LAN): A system of hardware and software designed to allow a group of intelligent devices to communicate within a fairly close proximity.

Lockout: The placement of a lockout device on an energy-isolating device in accordance with an established procedure to ensure that the energy-isolating device and the equipment being controlled cannot be operated until the lockout device is removed.

Lockout device: A device that utilizes a positive means such as a lock, either key or combination type, to hold an energy-isolating device in the safe position and prevent the energizing of a machine or equipment.

LSB: Least significant bit.

Machine language: A control program reduced to binary form.

MAP (manufacturing automation protocol): A "standard" developed to make industrial devices communicate more easily; based on a seven-layer model of communications.

Master: The device on a network that controls communication traffic.

Master control relay (MCR): A hardwired relay that can be deenergized by any hardwired series-connected switch and is used to deenergize all devices.

Memory map: A drawing showing the areas, sizes, and uses of memory in a particular PLC.

Microsecond: One-millionth (0.000001) of a second.

Milliamp: One-thousandth (0.001) of an amp.

Millisecond: One thousandth (.001) of a second.

Mnemonic code: A specific set of symbols designated to represent instructions in a control program; usually an acronym made by combining the initial letters or parts of words.

MSB: Most significant bit.

NEMA (National Electrical Manufacturers Association): An organization that develops standards that define a product, process, or procedure.

Network: A system connected to devices or computers for communication purposes.

Node: A point on the network that allows access.

Noise: The unwanted electrical interference in a programmable controller or network caused by motors, coils, high voltages, and welders that can disrupt communications and control.

Nonretentive coil: A coil that turns off upon removal of applied power to the CPU.

Nonretentive timer: A timer that loses the time if it loses the input enable signal.

Nonvolatile memory: The memory in a controller that does not require power to retain its contents.

NOR: The logic gate that results in zero unless both inputs are zero.

NOT: The logic gate that results in the complement of the input.

Octal: The number system based on the number 8, utilizing numbers 0 through 7.

Off-delay timer: A type that is on immediately when it receives its input enable and turns off after it reaches its preset time.

Off-line programming: The programming occurs when the PLC is not attached to the actual device and can then be downloaded to the PLC.

On-delay timer: A timer that does not turn on until its time has reached the preset time value.

One-shot contact: A contact that is on for only one scan when activated.

Open loop: A system that has no feedback and no autocorrection.

Operating system: The fundamental software for a system that defines how it will store and transmit information.

Optical isolation: Technique used in I/O module design that provides logic separation from field levels.

OR: Logic gate that results in 1 unless both inputs are 0.

Parallel communication: A method of communications data that transfers on several wires simultaneously.

Parity: A bit used to check for data integrity during a data communication.

Peer-to-peer: The communication that occurs between similar devices, such as two PLCs communicating.

PID (proportional, integral, derivative) control: A control algorithm used to closely control processes such as temperature, mixture, position, and velocity. Proportional portion takes care of the magnitude of the error; integral takes care of small errors over time; derivative compensates for the rate of error change.

PLC: Programmable logic controller.

Programmable controller: A special-purpose computer programmed in ladder logic so that devices could easily interface with it.

Proportional gain: A device that considers the magnitude of error and attempts to correct a system.

Pulse modulated: Turning a light source on and off at a very high frequency. In sensors the sending unit pulse modulates the light source. The receiver only responds to that frequency.

PPR (pulses per revolution): The number of pulses an encoder produces in one revolution.

Quadrature: The situation in which two output channels are out of phase with each other by 90 degrees.

Rack: A PLC chassis in which modules are installed to meet the user's need.

Radio frequency (RF): A communication technology with a transmitter/receiver that can read or write to tags.

RAM (random access memory): The normally considered user memory.

Register: A storage area typically used to store bit states or values of items such as timers and counters.

Repeatability: The ability to repeat movements or readings; for a robot, how accurately it returns to a position time after time.

Resolution: A measure of how closely a device can measure or divide a quantity; for example, an encoder resolution is defined as counts per turn.

Retentive coil: A coil that will remain in its last state even though power was removed.

Retentive timer: A timer that retains the present count even if the input enable signal is lost.

Retroreflective: A photosensor that sends out a light that is reflected from a reflector back to the receiver when the receiver and emitter are in the same housing.

RF: *See* radio frequency.

ROM (read-only memory): The operating system memory that is non-volatile and is not lost when the power is turned off.

RS-232: A common serial communications standard that specifies the purpose of each of 25 pins.

RS-422 and RS-423: The standards for two types of serial communication. RS-422 is a balanced serial mode. This means that the transmit and receive lines have their own common instead of sharing one like RS-232. Balanced mode is more noise immune. This allows for higher data transmission rates and longer transmission distances. RS-423 uses the unbalanced mode. Its speed and transmission distances are much greater than RS-232 but less than RS-422.

RS-449: An electrical standard for RS-422/RS-423 that is more complete than the RS-232 and specifies the connectors to be used.

RS-485: An electrical standard similar to RS-422 standard; its receivers have additional sensitivity that allows for longer distances and more communication drops.

Rung: A group of contacts that control one or more outputs; horizontal lines on the diagram in a ladder diagram.

Scan time: The amount of time for a programmable controller to evaluate a ladder diagram once; typically in the low-millisecond range.

SDLC: The Serial Data Link Control, subset of the HDLC, used in many communication systems such as Ethernet, ISDN, and BITBUS; defines the structure of the frames and the values of a number of specific fields in these frames.

Sensitivity: A device's ability to discriminate between levels.

Sensor: Normally a digital device used to detect change.

Sequencer: An instruction type device used to program a sequential operation.

Serial communication: A system of data represented by a coding system such as ASCII that is one bit at a time.

Slave: The nodes of the network that can transmit informations to the master only when they are polled (called) from it.

Speech module: A device used by a PLC to output spoken messages to operators.

Tagout: A device placed on an energy-isolating device, in accordance with an established procedure, to indicate that the energy isolating device and the equipment being controlled may not be operated until the tagout device is removed.

Tagout device: A prominent warning device, such as a tag and a means of attachment, that can be securely fastened to an energy-isolating device in accordance with an established procedure to indicate that the energy-isolating device and the equipment being controlled may not be operated until the tagout device is removed.

Thermocouple: A sensing transducer that changes a temperature to a current to be measured and converted to a binary equivalent that the PLC can understand.

Thumbwheel: A device used by an operator to enter a number between 0 and 9.

Timer: A device that changes its output state when a certain value is achieved.

TOP (technical and office protocol): A communication standard developed by Boeing that refers to the office and technical areas.

Transducer: A device that changes one form of energy to another.

Transitional contact: A contact that changes state for one scan when activated.

True: The enabling logic state generally associated with a "1" or a "high" state.

UL (Underwriters Laboratory): The organization that operates laboratories to investigate systems with respect to safety.

User memory: A device to store user information such as the user's program, timer/counter values, input/output status, and so on.

Volatile memory: The memory that is lost when power is lost.

Watchdog timer: A timer that can be set to shut the system down if the time is exceeded.

Word: Length of data in bits (16 or 32) that a microprocessor can handle.

Appendix A: Common I/O Device Symbols

AIR PREPARATION UNITS

Symbol	Description
	FILTER/SEPARATOR with manual drain
	FILTER/SEPARATOR with automatic drain
	OIL REMOVAL FILTER
	AUTOMATIC DRAIN
	LUBRICATOR less drain
	LUBRICATOR with manual drain
	LUBRICATOR with automatic filling
	AIR LINE PRESSURE REGULATOR adjustable, relieving
	AIR LINE PRESSURE REGULATOR pilot controlled, relieving
	FILTER/REGULATOR (piggyback) Manual Drain Relieving (With Gauge)
	FILTER/REGULATOR (piggyback) Auto Drain Relieving
	AIR LINE COMBO F-R-L simplified

PNEUMATIC VALVES

Symbol	Description
	CHECK
	FLOW CONTROL
	RELIEF VALVE

PNEUMATIC VALVES (Cont'd)

Symbol	Description
	2-POSITION 2-WAY
	2-POSITION 3-WAY
	2-POSITION 4-WAY
	2-POSITION, 4-WAY 5-PORTED
	3-POSITION, 4-WAY ports closed, center pos.
	3-POSITION, 4-WAY 5-PORTED cyl. ports open to exhaust in center position
	QUICK EXHAUST
	SHUTTLE

VALVE ACTUATORS

Symbol	Description
	MANUAL general symbol
	PUSH BUTTON
	LEVER
	PEDAL OR TREADLE
	MECHANICAL cam, toggle, etc.
	SPRING
	DETENT line indicates which detent is in use

VALVE ACTUATORS (Cont'd)

Symbol	Description
	SOLENOID
	INTERNAL PILOT SUPPLY
complete simplified	REMOTE PILOT SUPPLY
	AND/OR COMPOSITE solenoid and pilot or manual override
	AND/OR COMPOSITE solenoid and pilot or manual override and pilot

LINES AND FUNCTIONS

Symbol	Description
——————	solid line–MAIN LINE
- - - - - -	dashed line–PILOT LINE
··············	dotted line–EXHAUST OR DRAIN LINE
— · — · —	center line– ENCLOSURE OUTLINE
	LINES CROSSING (90° intersection not necessary)
	LINES JOINING (90° intersection not necessary)
	LINES JOINING
	FLOW DIRECTION hydraulic medium
	FLOW DIRECTION gaseous medium
●●●●●●●●●●●●●●●	ENERGY SOURCE
	LINE WITH FIXED RESTRICTION
	LINE WITH ADJUSTABLE RESTRICTION
	FLEXIBLE LINE
	PLUGGED PORT, TEST STATION, POWER TAKE-OFF
connected / disconnected	QUICK DISCONNECT WITHOUT CHECKS
connected / disconnected	QUICK DISCONNECT WITH CHECKS
connected / disconnected	QUICK DISCONNECT WITH ONE CHECK

401

Index